ENGINEERING GEOLOGY

MAXWELL PRESS

ENGINEERING GEOLOGY

W. Henry Penning F.G.S.,

Geologist, H.M. Geological Survey of England and Wales. Author of 'Field Geology,' etc.

MAXWELL PRESS

Chennai Trichy New Delhi

MAXWELL PRESS

This edition has been published in india by arrangement with Carson Books, UK

ISBN 978-93-5528-198-2

Maxwell Press
No. 44, Nallathambi Street,
Triplicane, Chennai 600 005

Printed and bound in India

MJP 1280

Publisher : C. Janarthanan

Publisher's Note

The legacy of a country is in its varied cultural heritage, historical literature, developments in the field of economy and science. The top nations in the world are competing in the field of science, economy and literature. This vast legacy has to be conserved and documented so that it can be bestowed to the future generation. The knowledge of this legacy is slowly getting perished in the present generation due to lack of documentation.

Keeping this in mind, the concern with retrospective acquiring of rare books has been accented recently by the burgeoning reprint industry. Maxwell Press is gratified to retrieve the rare collections with a view to bring back those books that were landmarks in their time.

In this effort, a series of rare books would be republished under the banner, "Maxwell Press". The books in the reprint series have been carefully selected for their contemporary usefulness as well as their historical importance within the intellectual. We reconstruct the book with slight enhancements made for better presentation, without affecting the contents of the original edition.

Most of the works selected for republishing covers a huge range of subjects, from history to anthropology. We believe this reprint edition will be a service to the numerous researchers and practitioners active in this fascinating field. We allow readers to experience the wonder of peering into a scholarly work of the highest order and seminal significance.

MAXWELL PRESS

PREFACE.

THIS work first appeared, during 1879, in the pages of *The Engineer* as a series of articles upon Engineering Geology, which are now reproduced—slightly altered in form, considerably enlarged, and more fully illustrated.

Engineering and Geology are so evidently and so intimately related that a knowledge of the former must include, and is incomplete without, an acquaintance with the latter; in turn, Geology derives much aid from engineering works, records and researches.

It is as an art that Geology must be treated for its results to be of immediate practical value to engineers; but as all art is based upon definite laws or principles, he will derive most benefit from Geology, and be the most proficient in its practical application, who founds his work upon it, also, as a science.

W. HENRY PENNING.

GRANVILLE HOUSE,
FINSBURY PARK, LONDON.

January, 1880.

CONTENTS.

PART I.

GEOLOGICAL STRATA, THEIR NATURE AND RELATIONS, AND THEIR BEARING UPON PRACTICAL WORKS.

CHAPTER I.

CHAPTER II.

GEOLOGICAL STRATA.

CHAPTER III.

GEOLOGICAL STRATA (*continued*).

CHAPTER IV.

GEOLOGICAL STRATA (*continued*).

PART II.

PROCEDURE IN THE FIELD.

CHAPTER I.

METHODS EMPLOYED IN GEOLOGICAL SURVEYING, MAPS AND SECTIONS.

CHAPTER II.

METHODS EMPLOYED IN GEOLOGICAL SURVEYING (*continued*).

CHAPTER III.

METHODS EMPLOYED IN GEOLOGICAL SURVEYING (*continued*).

CHAPTER IV.

METHODS EMPLOYED IN GEOLOGICAL SURVEYING (*continued*).

CHAPTER V.

METHODS EMPLOYED IN GEOLOGICAL SURVEYING (*continued*).

PART III.

ECONOMICS—MATERIALS, MINERALS AND METALS; SPRINGS AND WATER-SUPPLY.

CHAPTER I.

MATERIALS, MINERALS AND METALS.

CHAPTER II.

MATERIALS, MINERALS AND METALS (*continued*).

CHAPTER III.

MATERIALS, MINERALS AND METALS (*continued*).

CHAPTER IV.

SPRINGS AND WATER-SUPPLY.

CHAPTER V.

SPRINGS AND WATER-SUPPLY (*continued*).

CHAPTER VI.

LIST OF ILLUSTRATIONS.

ENGINEERING GEOLOGY.

PART I.

GEOLOGICAL STRATA, THEIR NATURE AND RELATIONS, AND THEIR BEARING UPON PRACTICAL WORKS.

CHAPTER I.

INTRODUCTION.

In the execution of engineering works, however scientific in design and clever in workmanship, failure has frequently usurped the place of success, because due attention has not been paid to geological phenomena. Numberless instances might be quoted in proof of this proposition, whilst it is notorious that vast sums of money have been thrown away in mining speculations which would at once have been characterised as hopeless by anyone possessing the slightest acquaintance with the science of Geology. A late eminent authority (Professor Jukes) has stated his belief that the amount of money fruitlessly expended in a ridiculous search after coal, even within his own experience, would have paid the entire cost of the Government Geological Survey of the United Kingdom.

Although a knowledge of this science is undoubtedly a great acquisition, which affords both pleasure and profit to its possessor, it is not possible, nor even desirable, for all professional engineers to become proficient geologists. Those for whom this work is more especially intended have too many claims upon their time and attention to bestow either upon a study of abstract principles, laws, and theories, which do not relate to their own particular science, art, or occupation; but they may nevertheless, and with advantage, avail themselves of the labours of others, when the results of those labours bear directly, and in a very important degree, upon the stability or success of the works designed or executed by them, or under their superintendence.

The engineer should certainly make himself acquainted with the geology of a district through which a railway is to be constructed from his designs and along the line of his selection. He should ascertain the nature of the various rocks that will be met with, not only at and near the surface of the ground, but for a considerable distance below; their relation to each other, and the important influence they will exert upon the works in contemplation. 'Trial-holes' are generally dug for this purpose, but these are simply pits excavated to a depth of a few feet, and afford information which, although valuable in itself, unless amplified in a particular manner, extends only to the superficial deposits. Deep borings are sometimes made, but are, in most cases, too costly; and however numerous these, or trial-holes, may be, both fall far short of what can be achieved in the same direction through the methods employed by the field-geologist. By these are determined, not only the kind of rocks occurring at or near the surface, but also their position in regard to each other; and the geological surveyor is enabled to indicate with reasonable accuracy what strata will be met with to a depth, it may be, of several hundred feet, and, what is of equal importance, the order of their succession. These results of his labours include not merely a knowledge of what beds would be pierced in sinking a well, or in excavating trenches for foundations, such as would be afforded equally by trial-holes or borings of sufficient depth; but they embrace also the important points of the 'lie' of the beds, the order of their superposition, their outcrop, dip, and consequent water-bearing properties; by all of which the stability and durability of engineering works are greatly affected.

We cannot fail to perceive how differently placed or constructed would have been many of the most important works, such as fortifications, railway-cuttings, embankments, tunnels, and even sewers, had those who designed them been acquainted with the principles, methods, and results of field-geology; or how much capital might have been usefully instead of fruitlessly expended, or how many catastrophes would have been averted. Mention has already been made of costly sinkings for minerals, where they could not possibly have been found; large sums of money have also been wasted in equally fruitless searches for water. Yet water-supply is as amenable to known laws as any other phenomenon of nature, and within certain limits it may be determined without experiment. Although the divining-rod has not even yet quite ceased to be a power amongst us, its days are surely numbered; men must, sooner or later, come to see that springs are merely the result of water finding its own level, and that for water to issue forth at one part of the earth's surface, it must have been absorbed at another. When the conditions affecting its absorption by and passage through the strata of a district are known or can be discovered, the existence of springs, their depth from the surface, and the height to which they will ascend, can be approximately, if not with extreme accuracy, determined.

In his 'Rudimentary Geology,' Major-General Portlock has truly and eloquently said: 'Geology is now a true science, being founded on facts and reduced to the dominion of definite laws, and in consequence has become a sure guide to the practical man. The miner finds in it a torch to guide him in his subterranean passage, to the stratum where he may expect to find coal or iron, or to the recovery of the mineral vein which he has suddenly lost; the engineer is guided by it in tracing out his roads or canals, as it tells him at once the firmest stratum for supporting the one, and the easiest to cut through for the other, and makes him acquainted with the qualities of the materials he should use in his constructions, and the localities where he should seek them; the geographer finds his inquiries facilitated by learning from geology the influence of the mineral masses on the form and magnitude of the mountains and valleys, and on the course of rivers; the agriculturist is taught the influence of the mineral strata on vegetable and animal life, and the statesman discovers in the effects of that influence a force which stimulates or retards population; the soldier also may find in geology a most valuable guide in tracing his lines both of attack and defence; and it is thus that a science rich in the highest objects of philosophical research is at the same time capable of the widest and most practical application.'

In the following pages, rules and methods relating to stratigraphical geology only are given, as the geological conditions which affect engineering and similar works are, mainly, the extent of the various strata, their lithological character, and their order of succession. It matters not what may have been the forms of Life during the ages when the strata were deposited, what their relations to those older or more recent, or what the order of their appearance in time; although the evidence regarding these points is as strong and as interesting as any upon which is based the science of Geology. The rocks are treated merely as stones, clays, and sands of varying kinds; some possessing commercial value and great utility; others having qualities to be guarded against in all mechanical operations; some only exhibiting water-bearing properties; but all worthy of study, independently of the old-world histories which they contain.

The names of places are given only in particular instances, such as those of mines, important quarries, notable sections, and so on, it having been considered advisable not otherwise to refer to localities in the description of the rocks. These are mentioned generally, and under specific denominations, geological maps indicating much more readily the formation at any particular spot than a lengthy reference to the many places which must otherwise have been mentioned as situated on an extended outcrop. Such maps are generally too small for the boundaries of formations to be defined upon them with extreme accuracy; indeed, they are not intended for that purpose, but rather to indicate what set, or sets, of beds an observer may expect to find in any particular neighbourhood. When an accurate delineation is required for any special purpose it may be found upon the sheets (corresponding in size and index numbers with the Ordnance maps) published by the Government Geological Survey. These are laid down upon a scale of one inch to a mile for the country generally, and for some districts upon the large scale of six inches to a mile; their prices vary, of the former (some of which may be obtained in quarter-sheets) from 4s. to 8s. 6d.; of the latter from 4s. to 6s. (p. 17). The main object of this work is, however, to enable the engineer to discover and, for all practical purposes, to trace out for himself the nature and extent of the rocks with which he is concerned.

An acquaintance with the methods of geological surveying is the more valuable, because 'drifts' are usually omitted from the maps; these are a series of superficial deposits which, although important in some localities, are not shown on any of the older geological charts, and are noticed on only a few of the more recent official publications. They consist chiefly of clays and gravels of peculiar character, which are found here and there upon the older rocks, on hills and in valleys, with no very definite mode of occurrence; and although, as a rule, of no great thickness, they of necessity exert a considerable influence over all works constructed upon them. A section is devoted to a brief description of these deposits, with the methods of tracing and mapping them, as they must also be treated from a practical point of view, in the same way as the older and more generally impor- formations.

CHAPTER II.

GEOLOGICAL STRATA.

Geological Strata.—Nature of the Rocks.—Relations of the Rocks.—Bearing of the nature of the Rocks upon practical works—*Railways—Tunnels—Embankments—Bridges.*

Geological Strata.—The crust of the earth consists of a great number of alternating rocky layers, various in kind, thickness, and extent, but always in regular, if not in constant, sequence. The uppermost have been formed in a great measure from the waste of those beneath, in the same way as the material now being deposited on the bottom of the sea has been derived from the denudation of the present dry land. These layers are but rarely horizontal, and they bear evidence of having been subjected to some upheaving force which has acted at various times, unequally and with different degrees of intensity, beneath every portion of the earth's surface. There have been corresponding, and on the whole nearly equal, movements of depression, and all areas have frequently been dry land, again to be covered by the waters of the ocean. It is owing to this inequality in the upheaval of the beds, and to their consequent partial destruction by the sea, that the lower and older strata are now exposed at the surface of the ground, and that we are enabled to classify the rocks and to decipher their ancient history.

The formations, of which the denuded edges are thus bared and thrown open to our inspection, are indicated by different tints upon geological maps. If it be borne in mind that each of the areas thus distinguished represents, as a general rule, the edge and not the surface of a formation, the proper apprehension of such maps is much facilitated. It is evident that were the variously-coloured portions each indicative of an original surface-plane, the rocks so depicted would generally be the newest, as overlying those which are hidden beneath. But their edges only being exposed and portrayed on the map, the main planes of bedding must now be either in a vertical position, or inclined from the surface in some direction, and the rocks, as a matter of course, must pass in under some of those that are contiguous. Geological maps show that, in this country, by far the larger proportion of the edges, or lines of outcrop, of the rocks, follow a nearly north and south direction, therefore the beds must dip, if at all, either to the east or to the west. The general dip of the rocks in these islands is, on the whole, towards the south-east; consequently those on the north-west are the oldest, and the lowest in the geological scale; those on the south-east are the highest in the scale, therefore the most modern.

All the beds of which the various geological formations are composed are termed 'rocks,' whether they are hard or soft, of aqueous or of igneous origin. The following remarks have been as far as practicable classified under three headings—(a) The nature of the rocks; (b) The relation of the rocks to each other; (c) The bearing of the nature and relation of the rocks upon practical operations.

(a) *The Nature of the Rocks.*—The aqueous and igneous deposits by which the known crust of the earth has been built up, occur in successive layers, and are of infinite variety as regards texture, colour, hardness, and other peculiarities. All the rocks are made up, wholly or in part, of minerals either in a crystalline or fragmentary form, or of mineral matter in a state of comminution. Some rocks contain metals, either in a free or native state, or, as ores, in combination with oxygen or sulphuric acid, whilst comparatively few are without metallic colouration.

All rocks may be divided into two great classes:—

1. The igneous, or unstratified, which (formed below the surface) were by volcanic or some similar force erupted through or intruded into the pre-existing formations. These are granites, traps and similar rocks.

2. The aqueous, or stratified, which were deposited from water as sediment, or (in some cases) as a chemical precipitate. They are chiefly clays, sandstones, limestones, and gravels.

There are rocks which have been otherwise formed, and some which have been altered from their original condition by heat or pressure, or by both agencies combined. Such metamorphic rocks may have been either igneous or aqueous, but are principally of aqueous origin, and are now found as gneiss, quartzites, marbles, slates, schists, and altered ashes.

The class to which a rock belongs is practically important, on account of the difference in the normal modes of its occurrence. The stratified rocks lie evenly, the one upon the other, whether horizontally or not, and preserve a regular but sometimes interrupted sequence; the unstratified follow no such definite lines, but are found suddenly breaking through older rocks and disappearing in an equally abrupt manner. In both classes the rocks of every kind present many varieties and gradations towards each other, but on the whole they possess broad characteristics by which they may be fairly determined. (See Some chapters on Lithology, and Tables for the determination of rocks, in 'Field Geology' (Baillière), from which those in Part ii. have been abridged.)

It may be noted that generally, but not without exceptions, the older stratified and the altered rocks are more crystalline and compact than are those of more recent date. Those that were by an old classification designated Primary, consist of slaty and crystalline strata, such as gneiss, and mica-schist, marble, and clay-slate; Transition, of slaty and siliceous sandstones and calcareous shales; Secondary, of chalk, limestones, red sandstones, marls, and clays; Tertiary, of sands and clays; Recent, of sands, gravels, silt, peat, and alluvium. The loose and friable beds are the most recent, overlying others more consolidated of secondary age, which in turn rest upon the more crystalline primary strata. All were once in the same unsolidified condition, but some have become hardened by chemical change and by the mechanical results of pressure and infiltration, during the ages which have elapsed since the time of their accumulation.

1. The more common igneous rocks are :—

GRANITE. SYENITE.	Granitic.
GABBRO.—'Greenstone.' FELSITE.	Trappean.
DOLERITE.—Basalt. PORPHYRITE. PHONOLITE. TRACHYTE.	Volcanic.

2. The aqueous rocks are :—

Argillaceous.

CLAY.
SHALE.—Hardened clay.
LOAM.—Clay and sand, a mechanical admixture.
LIMESTONE—when containing silicate of alumina: if this be in sufficient proportion it constitutes an hydraulic limestone.

Arenaceous.

SAND.
SANDSTONE.—Consolidated sand, with siliceous, ferruginous, or calcareous cementing material.
GRIT.—Coarse sandstone.
GRAVEL.
CONGLOMERATE.—Consolidated gravel.

Calcareous.

LIMESTONE.—Sometimes earthy as Chalk, oolitic as Bath freestone, crystalline as Marble.

MAGNESIAN LIMESTONE.—Limestone composed of carbonate of lime and magnesia.

SILICEOUS LIMESTONE.—Limestone containing much silica, as in Kentish Rag.

The altered or metamorphic rocks are :—

GNEISS.—A foliated rock, otherwise similar to granite in composition.

SCHIST.—Sedimentary rock, altered and foliated.

QUARTZITE.—Altered quartz sandstone.

SLATE.—Clay altered and cleaved in a direction generally transverse to its original bedding.

There are many other rocks in each class, also many which partake more or less of the character of each, presenting infinite gradations; but they occupy comparatively small areas, and in other respects exert the same influence as the rocks to which they are most nearly allied, they may therefore, from an engineering point of view, be considered as of slight importance.

(*b*). *The relation of the Rocks to each other.*—The relation of the rocks of a district—that is, their position in regard to each other, their relative thickness, dip, permeability, and so on—is quite as important for mechanical purposes as their individual nature. But this relation, especially in a complicated area, is not by any means readily ascertained, unless the proper methods of procedure be understood. The thickness of each necessarily rules the extent of ground it occupies, but must be studied in connection with the dip, which exercises also an influence on the shape of the country quite as powerful as that of the nature of the rocks of which it is composed.

Where the bedding of rocks is horizontal, or nearly so, the surface will be much more flat and spreading than where the dip is sharp, a condition which will produce a rugged and rapidly-alternating landscape. This fact is well worthy of notice, because we may reason conversely that if a country be flat, the local beds are tolerably level, and extend some distance in any direction; but if it be much broken, that they rapidly disappear, having a high angle of inclination. Upon the dip other properties of the beds depend; and it will be seen that it affects both directly and indirectly the works constructed on their outcrop. The relative elevation of varying deposits bears directly upon the flow of surface water from one area to another; therefore it affects the land springs, and in the same degree the dryness or dampness of any given locality. The question of relative permeability is more extensive and intricate, and upon this depend the all-important points of the power of absorption of water by the beds, and the nature and origin of deep-seated springs. These points influence not merely the supply to Artesian wells, but the liability to landslips, and must be considered also in calculating the varying pressures by which engineering works are especially affected.

The phenomena of deep-seated springs, just referred

to, depend not altogether upon permeability—although this is one of the chief elements in their production—but also upon the relative position of pervious and impervious strata. These may succeed each other in the simplest way, by being in regular sequence, with the higher beds resting evenly upon the lower, each possessing the same angle of dip; or in a more complicated manner, which is described as 'unconformable.' This term is applied to beds, or to sets of beds, which at any particular spot rest one upon the other, but possess different degrees of inclination. It is evident that in such a case the uppermost beds rest upon the edges, and not upon the surface of those beneath, and that before the higher were deposited, the lower had been cut off by some process of denudation. Occasionally beds overlap each other, without being exactly unconformable; and sometimes those which are known to be so, do nevertheless rest evenly upon each other, and with the same dip; but this is a merely local, and may be considered an accidental, occurrence.

When strata are unconformable, of course the continuity of the beds beneath the surface is broken, and this must affect, more or less, the flow of water through them in any direction. The underground extension of rocks is likely to be interrupted also by 'faults,' or other dislocations, by which portions of them are displaced, sometimes several hundred feet, and which may extend horizontally for a short distance only, or for several miles. It is necessary that all such faults or breaks be discovered, and their influence estimated, in the consideration of problems in engineering geology.

The relations of the rocks upon a grand scale, that is classified into 'formations' and 'series,' are shown upon good geological maps with as much minuteness as is possible upon the scale employed. The one-inch maps made and published by H.M. Geological Survey afford all needful information for general purposes, whilst the six-inch maps admit of greater detail and accuracy; but the former are not yet published for the whole country, nor are the latter for more than a comparatively small, although an important, portion. And there are frequent and sometimes rapid changes in the nature of the beds, which cannot be shown on maps that portray generally the structure of a district, although the local beds may be, and usually are, described in the memoirs by which many of the official maps are accompanied.

These local peculiarities render an acquaintance with the methods of geological surveying valuable to the engineer, who would thereby be enabled to discover and to trace out for himself all local changes and intercalated deposits by which his works may be greatly affected. For instance, in many thick clay formations, shown wholly as such on the maps, there are limited beds of limestone, sometimes in considerable number and of excellent quality, which may be utilised for building, for lime, or for ballast; on the other hand, there are in similar positions, intercalated deposits of permeable material which yield water that, in some cases, must be guarded against, in others, may be turned to good account.

The following list of the maps officially published by the Geological Survey may be found useful for reference, the numbers of the sheets being precisely the same as those of the Ordnance Survey.

LIST OF MAPS,

PUBLISHED BY H.M. GEOLOGICAL SURVEY.

SCALE, one inch to a mile.

ENGLAND AND WALES.

			s.	d.
	Large sheet, London and its Environs	-	22	0
	„ „ „ „ with Drift deposits		30	0
	Sheets, representing an area of about 800 miles.			
No. 1.	In quarter sheets	- each quarter	3	0
2.	Whole sheet	- - - -	4	0
3.	„ „	- - - -	8	6
4.	„ „	- - - -	5	0
5 *and* 6.	„ „	- - - each	8	6
7.	„ „	8*s*. 6*d*. With Drift deposits	18	6
8 *and* 9.	„ „	- - - each	8	6
10.	„ „	- - - -	4	0
11 *to* 22	„ „	- - - each	8	6
23 *and* 24.	„ „	- - - „	4	0
25 *and* 26.	„ „	- - - „	8	6
27 *to* 29.	„ „	- - - „	4	0
30 *and* 31.	„ „	- - - „	8	6
32.	„ „	- - - -	4	0
33 *to* 37.	„ „	- - - „	8	6
38 *and* 39.	„ „	- - - „	4	0
40 *and* 41.	„ „	- - - „	8	6
42 *and* 43.	In quarter sheets	- each quarter	3	0
44.	Whole sheet	- - - -	8	6
45 *and* 46.	In quarter sheets	- „ „	3	0
48.	S.E. Quarter sheet, 3*s*. With Drift deposits	- - - -	3	0
52 *to* 56.	In quarter sheets	- each quarter	3	0
57.	N.E., S.W., and S.E. Quarter sheets	each	3	0
57.	N.W. Quarter sheet	- - -	1	6

			s.	d.
No. 58.	Whole sheet - - - -		4	0
59.	N.E. *and* S.E. Quarter sheets, each		3	0
60 *to* 63.	In quarter sheets - each quarter		3	0
64.	Whole sheet - - - -		8	6
71 *to* 75.	In quarter sheets -		3	0
76.	N. Quarter sheet - - -		1	6
76.	S. ,, ,, - - -		3	0
77.	N.E. ,, ,, - - -		1	6
78 *to* 82. / 87 *to* 89.	In quarter sheets -		3	0
90.	N.E. *and* S.E.	Quarter sheets, each quarter	3	0
91.	N.W. *and* S.W.			
92.	S.W. *and* S.E.			
93.	N.W. *and* S.W.			
98.	N.E., S.W. *and* S.E.			
101.	S.E.			
105.	In quarter sheets		3	0
109.	S.E. Quarter sheet - - -		3	0

Six-inch maps are published of parts of Northumberland, Cumberland, Durham, Lancashire, and Yorkshire, in sheets representing about 24 square miles, at 4*s.* or 6*s.* per sheet.

SCOTLAND.

			s.	d.
No. 1 *and* 2.	Whole sheets - -	each	4	0
3, 7 *and* 9.	,, ,,	- ,,	6	0
13.	- -	-	4	0
14 *and* 15. / 22, 23 *and* 24. / 32 *and* 33.	,, ,, - -	- ,,	6	0
34.	,,		4	0
40 *and* 41.	,, ,, - -	- ,,	6	0

Six inch maps are published of parts of Edinburghshire, Haddingtonshire, Fife, Ayrshire, Renfrewshire, Dumbartonshire, Lanarkshire, Dumfriesshire, Stirlingshire, Perthshire and Clackmannanshire, in sheets representing about 24 square miles, at 4*s.* or 6*s.* per sheet.

IRELAND.

			s.	d.
No. 21, 28, 29, 35, 36 *and* 37.	Whole sheets -	- each	3	0
38.	" " -	-	1	6
47, 48 *and* 49.	" " -	- "	3	0
50.	" " -	-	1	6
53, 59 *to* 61, 66 *to* 71.	" " -	- "	3	0
72.	" " -	-	1	6
74 *to* 81.	" " -	- "	3	0
82.	" " -	-	1	6
83 *to* 121.	" " -	- "	3	0
122.	" " -	-	1	6
123 *to* 130.	" " -	- "	3	0
131.	" " -	-	1	6
132 *to* 134.	" " -	- "	3	0
135.	" " -	-	1	0
136 *to* 139.	" " -	- "	3	0
140.	" " -	-	1	6
141 *to* 149.	" " -	- "	3	0
150.	" " -	-	1	6
151 *to* 158.	" " -	- "	3	0
159 *and* 160.	" " -	- "	1	6
161 *to* 169.	" " -	- "	3	0
170.	" " -	-	1	6
171 *to* 179.	" " -	- "	3	0
180 *to* 182.	" " -	- "	1	6
183 *to* 188.	" " -	- "	3	0
189 *and* 190.	" " -	- "	1	6
191 *to* 195.	" " -	- "	3	0
196 *and* 197.	" " -	- "	1	6
198 *to* 201.	" " -	- "	3	0
202 *to* 205.	" " -	- "	1	6

Six-inch maps are published of parts of the counties of Roscommon, Leitrim, Sligo, Antrim and Tyrone, in sheets representing about 24 square miles, at 4*s*. or 6*s*. per sheet.

Geological Sections, on a natural scale of six inches to a mile, traversing England and Wales, Scotland and Ireland, in various directions, are also issued, with descriptions, in sheets representing about 36 lineal miles, numbered 1 to 121, at 5*s.* per sheet.

There are also Vertical Sections, on a scale of 40 feet to an inch, through Coal Measures and other important strata, in sheets numbered 1 to 62, at 3*s.* 6*d.* per sheet.

Memoirs on separate areas, and on various subjects in Physical Science, and Records of the School of Mines and of Science applied to the Arts, are also published by the Geological Survey. (For full particulars of all these works and of the various maps and sections, see the 'Catalogue of the Publications of the Geological Survey of the United Kingdom.')

(*c*) *The Bearing of the Nature* (*a*) *and Relation* (*b*) *of the Rocks upon Practical Works.*—The nature of the rocks which form the surface of any district is indicated generally upon geological maps of the area in question, and such maps are of great value, to the engineer, as the rocks, whatever their nature may be in any particular locality, of course form the base of all engineering and also of all architectural works. The same remark applies equally to mining and well-boring operations, and to agricultural pursuits; the rocks must, of necessity, exercise over them all a permanent influence, greatly conducive either to their success or to their failure. The rocks exert this influence not only by their inherent properties or peculiarities, but also, by the relations which they bear towards one another. These relations are shown in the sections, which amplify the information supplied by the maps; they have been briefly mentioned above, and are more fully described further on. Some of the many instances in which the influence of the rocks is exercised may be here enumerated, as indicating the necessity for all available geological information being secured, and calculations based thereupon, before works of any importance are commenced or even designed. This influence is exerted in two distinct ways, (*a*) Through the nature of the rocks; (*b*) Through their relation; but the two, although distinct, are sometimes blended, and always so intimately associated that a partial separation only can be attempted.

(*a*) THE BEARING OF THE NATURE OF THE ROCKS UPON PRACTICAL WORKS.

Cuttings.—The trial-holes usually made before the commencement of railways, and other engineering works, enable the engineer to determine the kinds of rock which come to the surface along a given line, or over a given area. But these holes, by themselves, afford no indication of the thickness of any great deposit, and in this respect may mislead, unless the actual conditions be otherwise ascertained; a knowledge of several particulars can, however, be derived from trial-holes, when once the proper method of utilising their indications is understood. For instance, a line of such holes traversing a hill may be, almost without exception, in clay, one or perhaps two of them on the flank of the hill being sunk in a hard rock. The inference would probably be that a cutting made through this hill must pass entirely through clay, except at those points where the hard rock was actually exposed; the reality, that nearly the whole of the work has to be carried on through the harder stratum, perhaps at a great additional cost. For the edge only of the bed was touched by the one or two holes where it comes to the surface, with a narrow outcrop, along the steepest part of the hill.

This sort of thing has repeatedly happened, and more frequently still, a bed of gravel or sand, only a few feet in thickness, but spread over an extended area or on the slope of a hill, has led to the conclusion that the whole of the cutting would be, as all the trial-holes were, in sand or gravel. It may turn out to be hard and intractable rock, removable only by blasting operations, but covered with gravel just thick enough to reach below the few feet exposed in the trial-holes. Or the converse may be the case; the engineer unacquainted with geological methods feels sure, from his trial-holes, that he will obtain from a certain railway cutting, building-stone sufficient for all the bridges, or gravel enough to ballast the line; but as the work progresses, and the deeper strata come into view, he meets with serious disappointment. These instances, selected from many, are sufficient to show the necessity for the exact nature of the rocks being ascertained, not merely at the surface, but to a depth below, certainly not less than that of the deepest cuttings. (See also *Banks*, p. 25.)

The nature of the rocks to be passed through in cuttings materially affects the preliminary calculations in regard to the slopes and necessary widths, which differ accordingly, for the 'angles of repose' vary directly as the nature of the rocks in question. It will be seen from the following table that these angles cover a wide range, from the very flat slope of 14° up to a vertical face, if the harder rocks not named in the table be taken into consideration.

TABLE OF NATURAL SLOPES OF EARTHS,

WITH REFERENCE TO A HORIZONTAL LINE.

	*Molesworth.**	*Rankine.*†
Clay, wet	16°	14° to 17°
Sand „	22°	21° to 37°
Vegetable soil	28°	
Sand, dry	38°	
Shingle	39°	35° to 48°
Gravel	40°	
Rubble	45°	
Clay, well drained	45°	
Compact earth	50°	

Chalk, when compact, will stand in cuttings with a vertical face, as will the tougher and more finely grained varieties of sand when closely bedded; indeed a series of narrow benchings, with vertical faces, is to be preferred to a uniform slope in such material. The slopes of the banks formed of the excavated material will, of course, be the same as those given in the above tables.

The hard rocks, such as limestone and sandstone, will stand safely in cuttings at angles varying according to the number and extent of joint- and bedding-planes; when these are few and narrow the face may be vertical, being sloped or benched back as they become more numerous or extensive. In such cases the dip of the beds must be considered; a face may

* 'Pocket-book of Engineering Formulæ.'
† 'Manual of Civil Engineering.'

be vertical on the lower side of a cutting as regards the dip with perfect safety, but on the upper side may have to be cut back very considerably, especially if the material between the beds of rock be at all of a yielding or slippery nature. (See also p. 45.)

Professor W. J. M. Rankine, in his 'Manual of Civil Engineering' (1864, p. 317), makes the following remarks on this important subject :—

'When rock is firm and sound, so that the permanence of its cohesion may be depended upon, the sides of excavations in it may be made vertical, or nearly so.

'How far the cohesion of the rock is to be depended upon, is a question to be solved rather by observation of the rock in each particular case, than by any geological principles having regard to its geological position, mineralogical character, or chemical composition; for the geological position is fixed by the organic remains imbedded in the rock; and these have no connection with its mechanical properties; and rocks composed of the same species of minerals, and the same chemical constituents in the same or nearly the same proportions, show great differences in strength and durability.

'It may be observed, however, that the cohesion of igneous and metamorphic rocks, such as granite, syenite, trap, gneiss, mica-slate, marble, quartz-rock, etc., may in general be trusted, unless they are much fissured, or contain potash-felspar, in which cases a sufficient slope must be given, to prevent fragments from falling into the cutting. Of the sedimentary rocks, those which contain much clay, such as shale, are to be treated with caution, how hard soever they may be when first cut; for they are liable to soften by the action of the weather. Sandstone and limestone, whether compact or granular, if fit for building purposes, will stand with vertical or nearly vertical faces; but those materials exist of every degree of hardness, from that of rock, properly speaking, to that of earth. Sandstone is met with which crumbles in the hand, and requires slopes of from 1 to 1 to 1½ to 1; and chalk, according to its degree of hardness and soundness, stands at slopes varying from $\frac{1}{2}$ to 1 to $1\frac{1}{2}$ to 1.

'The stability of sedimentary rocks in the side of a cutting is greater when the beds are horizontal, or dip away from the cutting, than when they dip towards it.'

Tunnels.—If information regarding the nature of the rocks be necessary for cuttings through them, it is much more required for tunnels, where the work is far more costly, and where, consequently, much greater saving may be effected by a previous knowledge of what strata will, or will not, be passed through in any line at a given level. Yet here the indications from trial-holes must be still more meagre, seeing that tunnels are seldom made except where the hills are too lofty to be passed through by open cuttings, consequently through strata at a greater distance from the surface. But the evidence so obtained, if treated by geological methods, may be made equally reliable, and in proportion, far more valuable. Many tunnels have been made in places which would have been sedulously avoided had the geological phenomena been previously ascertained; others, where, by diverting the line a short distance to one side or the other, the tunnels might have been made with a saving of more than half the cost of construction.

Banks.—The nature of the rocks to be passed through in tunnels or cuttings affects also the calculations for the necessary slopes, and consequent widths, of embankments. For the 'angles of repose' vary in this case also, as the material, of which they are to be formed, is sand, gravel, clay, chalk, or rubble; the latter being the form which the harder rocks would assume after excavation. (See pp. 22, 23.)

Main Drains, Docks, etc.—In main-drainage works, trial-holes afford ample information regarding the kind of strata along which the sewers are to be laid, but not as to their relation—a much more important point in this particular class of work, and one which is referred to further on. The remark is applicable also to excavations for docks, foundations for dock walls and sills, water towers, and similar works. The evidences of the solidity of the rock, its inherent liability to slip or to squeeze outwards under pressure, being obtainable from trial-holes and borings, need not here be considered.

Bridges.—In many cases the nature of the rock upon which bridges or culverts are to be built can be very well ascertained by trial-holes; but not by any means in all. For it happens that many of the largest and most important structures, such as bridges and viaducts, are required in the lower-lying parts of a district, that is, in its valleys; and here the evidence thus obtained is apt to be misleading. The smaller valleys are, in places, filled to a depth of many feet with a wash from the neighbouring hills, composed perhaps of sand or clay; which wash so closely resembles the rock whence it has been derived, that it is, except to the experienced eye, a part of, and continuous with, the rock of which the high ground consists. But beneath this wash there may be a treacherous bed of peat, or even of quicksand, the evil influence of which is perhaps only discovered when the new structure is sufficiently advanced for its weight to cause an awkward settlement.

In such places, it may be urged, borings are resorted to rather than trial-holes; even where such is the case similar results may occur, should the misleading characters be repeated. And it often does happen that in alluvial flats there is a great number of rapidly alternating ancient river deposits, which may consist of peat, silt, gravel, sand, or clay. The solid substratum may not be reached perhaps for fifty, sixty, or even a hundred feet, if the spot in question be situated over an old course of the river, sometimes a long way from its present channel, and of which there is nothing on the marshy plain to indicate the existence.

CHAPTER III.

GEOLOGICAL STRATA—*continued.*

Bearing of the nature of the rocks upon practical works, *continued—Materials—Minerals—Metals—Agriculture—Land-drainage—Sewerage works.*

Materials.—A very important element in the cost of construction in all engineering and building works is the material which the rocks of the neighbourhood will yield, and this of course varies, both in quantity and quality, according to the nature of the rocks themselves. It may be assumed that the local quarries, lime-kilns, brick-yards, etc., will almost certainly be on the outcrop of beds most prolific in building materials, and afford good evidence of the kind required. But geological knowledge is nevertheless requisite to guide the engineer in laying out his works so that they may strike the more valuable strata to the best possible advantage.

Building material is frequently brought long distances, when that which is as good, or even better, occurs in the vicinity—although perhaps hidden by a few feet of drift—it may be in abundance. A railway-cutting or a tunnel may be judiciously set out so as to follow exactly the course of a useful stratum, even to a considerable depth from the surface, probably to rail level. On the other hand, it may be planned so as to miss the bed, except just at the surface, or possibly altogether; for the point in question depends upon the direction of the dip of the stratum, its consequent strike, and the actual amount of its inclination.

The drift gravels occur in a more irregular manner than any other series of deposits, but if they be previously mapped, and the work be designed accordingly, a great saving may be effected; the labour of excavating a cutting, for instance, may perhaps be made to yield the additional result of affording ballast or road-metalling. The extent and thickness of the gravels should be ascertained, also their mode of occurrence, whether as capping a ridge, filling an old channel, or resting on the sloping flank of a hill (see Part ii. chap. 4). Where gravels are scarce, or altogether absent, ballast may sometimes be obtained from the more solid rocks, broken up small for that purpose, and in some districts these are sufficiently plentiful. Even in many thick deposits of clay there are found occasional beds of hard septaria, or thin bands of limestone, suitable for the purpose; the line of these, when their outcrop has been traced, may frequently be followed with advantage.

Nearly, if not quite, all the geological formations yield some one or more forms of building material, and many consist almost entirely of rocks that can be utilised in construction. The varieties and qualities are numerous, still nearly all fall under the general terms granite, limestone, sandstone and brick-earth.

'Everyone who has carefully inspected a good collection of rock specimens must have noticed the great numbers there are, both of different kinds of rock and of intermediate varieties. Distinct as they may be in some particulars, these varieties are still so nearly alike, that a series may frequently be selected, presenting a perfect gradation between rocks that are distinct if studied by themselves, and partaking more or less of the characters of each.' ('Field Geology,' p. 152.)

There are many series of strata, in which the rocks, being either all limestones, or all sandstones, or a mixture of both, with perhaps intervening clays, are grouped under some comprehensive term. These general and inclusive denominations are convenient rather than strictly accurate; as instances may be mentioned the Lower Oolite limestone and the Caradoc sandstone, each series containing beds of a different character to that implied by their group names.

In every series there are some beds of more especial value for particular purposes than others above and below them, although the difference may not be at once apparent. There are beds also which for building works should be scrupulously avoided, in consequence of their possessing some detrimental peculiarity of composition. As beds vary rapidly, that one which is good in every respect in one district being worthless in another, no general description can accurately apply to all localities; therefore this point should receive careful and local investigation.

The granites are unstratified rocks, that is, they exhibit no lines of lamination which, in stratified rocks, mark the original layers of deposition; and they consist mainly of crystals of quartz, felspar, and either mica or hornblende, or they may present confused aggregates of those minerals in a more or less crystalline form. These rocks are split up by joints into irregular tabular masses, by which the work of quarrying is much facilitated.

'The durability and hardness of granite are the greater the more quartz and hornblende predominate, and the less the quantity of felspar and mica, which are the more weak and perishable ingredients. Smallness and lustre in the crystals of felspar indicate durability; largeness and dullness, the reverse. The best kind of granite are the strongest and most lasting of building stones.'*

The limestones are stratified rocks of infinite variety, generally occurring in regular series of beds, alternately with clays, but sometimes as isolated beds, or as lenticular masses, in the midst of other deposits. They may consist of nearly pure carbonate of lime, of a double carbonate of lime and magnesia, or with an admixture of sand or clay, when they are called arenaceous or argillaceous limestones, according to the nature of the material which predominates. Limestones are durable in proportion to their texture, those which are compact resisting the action of frost and the solvent power of the free carbonic acid in the air better than those which are coarse or porous.

Those limestones which are nearly pure are, as a rule, the best suited to the purpose of being burnt into quicklime, provided they are not too compact, in which case the additional fuel required adds greatly to the cost of calcination. An important exception is that kind of limestone which contains from ten to

* Rankine's 'Manual of Civil Engineering,' 1864, p. 356.

twenty-five per cent. of clay, and yields on burning a cement which sets under water, and is commonly known as hydraulic lime.

'The larger the proportion of clay in the stone, the more rapidly the cement becomes solid, the hardening being complete in two or three days when the proportion amounts to twenty-five per cent., and taking three weeks when only ten per cent. Much depends (especially in artificial admixtures) on the minute division and perfect admixture of the foreign particles.'[a]

The sandstones also are stratified rocks which occur under many different conditions of texture and composition. They consist of quartz grains, more or less rounded, compacted by pressure, and cemented together by siliceous, calcareous, or aluminous matter that has been deposited within them from water holding one or more of those substances in solution. Upon the nature of this material, as the grains are the same in all, the durability of the stone, of necessity, depends. The clayey sandstones are soft and perishable, the calcareous disintegrate rapidly when exposed to the weather, whilst those with a siliceous cement are by far the most durable. Sandstones should be selected for building purposes which are crystalline in texture, not dull and earthy-looking, and which do not readily separate along their lines of lamination.

'Sandstone is in general porous, and capable of absorbing much water; but it is comparatively little injured by moisture, unless when built with its layers set on edge, in which case the expansion of water in freezing between the layers makes them split or scale off from the face of the stone. When it is built

[a] Ansted's 'Geology,' 1856, p. 462.

on its natural bed, any water which may penetrate between the edges of the layers has room readily to expand or escape.

The better kinds of sandstone are the most generally useful of building stones, being strong and lasting, and at the same time easily cut, sawn, and dressed in every way, and fit alike for every purpose of masonry.'*

The brick-earths, as here considered, include all those clays which are, or may be, used for the manufacture of bricks, drain-pipes, etc. Brick-earth, of typical quality, should consist of an intimate mixture of pure clay with clean sand; the proportions of these ingredients may vary, but that of lime should not exceed two per cent., as it causes disintegration of the bricks; nor is the presence of too much iron desirable, or the bricks become vitrified in process of burning.

'Compounds of silica with one other earth are difficult of fusion, and resist the most intense heat of a furnace . . . such clays only are fit to make fire-bricks and crucibles, and to cement together the parts of furnaces.

'Stourbridge fire-clay consists of:

One equivalent of alumina,
Two equivalents of silica,

with two equivalents of water, and a small quantity of oxide of iron.

Double silicates are more easily fusible; common clays, by the presence of silicates of lime, magnesia, and protoxide of iron; and the bricks made of them, when thoroughly burned, are partially vitrified. Silicate of lime in the clay, in any considerable quantity, makes it too fusible, so that the bricks soften in the kilns and become distorted. Clay containing carbonate of lime should be avoided . . . sand in moderate quantity is beneficial; excess makes the bricks too brittle. One part by volume of sand to four or five of pure clay is about the best proportion.'†

* Rankine, *Op. cit.*, p. 358. † Rankine, *Op. cit.*, pp. 364-5.

The following table gives the weight of the more common building stones and other materials, with the force required to crush them; those which are the heaviest, of their own kind, being usually the strongest. Some stones absorb much more water than others of the same class, and are proportionately less durable; this peculiarity varies in granite from 1·4 to 12·5 parts per thousand, and in hard York sandstone from 16·5 to 33· parts per thousand.

TABLE OF WEIGHT AND RESISTANCE TO CRUSHING OF THE MORE COMMON ROCKS.

		Weight per cubic foot in lbs.		Crushing force in lbs. per square inch.
Granite		†164 to 172*	(Mount Sorrel)	12,861*
			(Argyllshire)	10,917*
Syenite			(Mount Sorrel)	11,820*
Basalt		187*		11,970*
Trap		170*		
Grauwacke				16,893*
Marble		168†		6,000†
Limestone		156†		3,000 to 8,000†
"		169 to 175*	(strong)	8,528*
"	Magnesian	178*	(strong)	7,098*
"	"		(weak)	3,050*
"	Bath Oolite	112†		
"	Portland Oolite	131†		3,700†
Chalk		143*		400†
"		117 to 174*		
Sandstone		156†		5,000†
"	(average)	144*		
"	(various)	130 to 157*		3,000 to 3,500*
Grit		131†		
York flagstone		143†	(mean)	9,824*
Clay		125†		
Shale		162*		
Slate		†175 to 180*		
Sand		120†		
Gravel		120†		
Shingle		90†		

* Rankine, *Op. cit.* † Molesworth, *Op. cit.*

Minerals and Metals.—The minerals and metals which occur so abundantly in Great Britain, and the cost of mining for them, are directly dependent upon the nature of the rocks with which they are associated, and the conditions by which those rocks have been affected since their deposition. In the most important instances they exist as an integral part of the formations in which they are found, as coal, ironstones, etc., but these are not, strictly speaking, either minerals or metals. Pure minerals and native metals are comparatively rare, but the terms are conveniently, if somewhat loosely, extended to include rock-masses, or portions of rock-masses, of which certain mineral matters, or metallic ores, form a characteristic part. The 'minerals' and 'metals,' included in the terms thus qualified, may be for all practical purposes treated in the same way as the rocks in which they are enclosed, or with which they are inter-stratified. Their existence in any given area can be ascertained in a similar manner, their outcrop surveyed, their extent determined, and their value approximately estimated.

Agriculture.—All agricultural pursuits are affected to a degree much greater than is perhaps generally understood, by the nature of the solid rocks beneath the surface soil on which the success of farming operations is universally admitted to depend. For all soil, or mould, has been produced, during the lapse of ages, by the atmospheric disintegration of the surface of those strata which form the base or subsoil. It has been increased in depth and somewhat modified, but its constituents have not been materially altered, by the annual growth and decay of vegetable matter thereon. The process has been aided by the ceaseless action of earth-worms working into and turning up the subsoil; ploughing has the same effect of increasing the depth of soil by constantly adding fresh material of similar character. It is evident that the nature of the soils of any district must therefore vary as the subsoils or strata from which they have been derived, and in a corresponding degree; and that the geology of a place being known, the nature of its subsoils and soils is at the same time determined.

Gravel and sand always produce light soils, abounding in silica, which may vary from fine sandy mould to stony soil, as the particles of the rock beneath are fine and uniform, or coarse and irregular. Clay forms stiff, heavy, and sometimes tenacious soil, varying in quality, but generally productive. Limestones produce light variable soils, sometimes full of detached lumps of the rock, and generally yielding good returns for high cultivation.

There are many physical characters which modify the normal characters of the soils derived from subaërial disintegration of the rocks; the results of the influence exerted by these causes may not be extensive, but they are locally important. First may be mentioned the rain-wash, which results from the lighter particles of the rocks being washed down by rain from higher to lower ground; accumulations of this material may often be seen several feet in thickness. Then, where the rain-wash has been arrested by a wall or fence on a hill-side, the result is, after a time, very evident, in the ground being higher on the upper than on the lower side. The growth of peat and the accumulation of marsh-clay are agencies which give rise to soils very different from what they would otherwise have been at the spots beneath such influences. Again, the soil of which a marshy plain is composed is due to the rocks somewhat further up the valley, rather than to the rocks actually beneath, the particles having been brought down in suspension by the waters and deposited, thus forming a flat, when the stream has at times overflowed its banks. Further, where two indifferent soils meet, which have been formed from decomposition of contiguous rocks, that which occurs along the line of junction is generally found to be of better quality, owing to admixture.

Some soils are rich in fattening properties and excellent for grazing, but through want of lime, without which no bone can be formed, young stock do not thrive upon their produce. The deficiency can often be supplied at a small cost, and the value of the land be thereby much enhanced; this is frequently the case in low marshy situations. Other soils, having had much of their productive properties removed by excessive croppings, may be renovated by a surface dressing of the parent subsoil; the same object may be effected by deeper ploughing, the subsoil being of course gradually incorporated with the surface mould. But for these and similar operations, certain particulars must be obtained, or the labour so expended will perhaps have been thrown away. These are the chemical composition of the subsoil, and the constituents required to be added to the soil itself; details which are readily obtainable through a small expenditure of time or money upon chemical, but not necessarily quantitative, analysis.

Land-drainage.—As the subsoil varies, so does the necessity for draining the land, and the facility with which the operation can be performed. Land-drainage, as ordinarily understood, is a simple matter, but there are some geological considerations respecting it to which attention may be briefly directed. As the strata affect the soils and sub-soils, so they must of necessity also exert an influence upon the natural drainage, and upon the means best adapted to that of an artificial character. The springs of one locality being but the natural outlet of water from another, the strata that now throw out springs would, if occurring at a different level, act as the channels for draining water away from the surface to be afterwards thrown out as springs elsewhere. And it sometimes happens that the subsoils, or underlying strata, may be, by some artificial aid, made available for purposes of drainage where they would not so act without that assistance. In other words, a plan of combined natural and artificial drainage can sometimes be easily carried out where a natural system does not exist, and where an entirely artificial scheme can be adopted only with very considerable trouble and expense.

Sewerage Works.—Although the last few years have witnessed a great, and, on the whole, beneficial change in the methods adopted for the disposal of town-sewage, that difficult problem is still far from being solved. In many cases plans of irrigation have been carried out, and these are always affected by the nature of the rocks upon which the sewage farms are situated. Opinion is greatly divided, not only as to the respective merits of the methods of precipitation and irrigation, but also, when the latter plan is in question, regarding the kind of soil best suited to the purpose.

A heavy soil will frequently yield enormous crops when judiciously irrigated, and in favourable seasons; but beyond the mechanical deposition of its suspended particles, the water is not clarified in the process; it runs off the land almost as chemically impure as when pumped or discharged from the reservoir. Very little of the liquid percolates down into a typical clay, and the benefit derived by the crops seems to be mainly owing to the moistening of the surface when it would otherwise be dry and parched; it is probable that pure water would have almost as good an effect as liquid sewage upon heavy land. On the other hand, light soils absorb, and for a while retain, a great deal of the moisture, giving it out again to the crops in a more equable manner; they may yield less produce, but this is, in some measure, through their containing within themselves a smaller proportion of the elements of fertility. It seems reasonable to suppose that as the liquid is to a great extent filtered by its passage over or through sandy soils, the ingredients removed from it are ready for absorption by vegetation. A light soil, owing to its permeability, will at all times take more sewage liquid, acre for acre, than a heavy one, and in some seasons—during a period of floods, for example—this property may prove of great advantage.

Doubtless much may be said on behalf both of the light and heavy soils, of the gravels and the clays, but a point that should not be lost sight of is the ultimate disposition of the water holding sewage particles in solution and suspension. For considering the phenomenon referred to in connection with land drainage, that the springs of one part are but the drainage of another, it is questionable how far we are justified in saturating the sands of any locality with tainted water. (See p. 38.)

There are some soils, such as loam, intermediate between sand and clay, combining the characters of both in proportion to the quantity of each which occurs in their composition; and this kind of soil may ultimately be found the best for sewage irrigation. Or, what is more probable, an area partly on impervions clay and partly on pervious sand or gravel will offer the greatest advantages. For in addition to being somewhat more independent of the seasons, farms so situated would admit of the water, after running over the clay and there depositing its suspended matter, passing, by gravitation if possible, on to the pervious beds. In its passage over or through these it would be more or less filtered, and the effluent water be thus rendered, perhaps, sufficiently pure to be allowed to flow into a stream or river with impunity.

CHAPTER IV.

GEOLOGICAL STRATA—*continued.*

Bearing of the relations of the rocks upon practical works—*Mining operations—Railway cuttings—Tunnels—Embankments—Reservoirs—Canals—Main-drainage—Foundations—Water-supply—Dampness—Disease.*

THE term 'relations of the rocks,' as used in this work, has reference not merely to the relative positions occupied by two or more strata, although that is the more important of their relations as generally understood. It includes the relative numbers of the beds or series of beds under investigation, as the proportionate number of rocks, possessing different characteristics in any area, exercises an important influence in all local questions of engineering geology. The thickness of each bed and of each series of beds, their constancy or variation, form also important elements, and with these may be associated relative surface elevation. Another is relative permeability (although this characteristic comes under the term the 'nature of a rock'); the relation of permeable and impermeable beds in alternation, in juxtaposition, or with beds of intermediate character between, greatly affects all practical works, when the occurrence of one only, of either kind, would call for no special investigation.

Perhaps the most important feature in the relation of rocks to each other is the amount of dip possessed by each, and its direction. The various and varying dips of the rocks must necessarily bring them in contact with each other under different conditions, and must produce folds and flexures amongst themselves, which give rise to anticlinal ridges and synclinal troughs or valleys. When all the rocks observable in a given area are horizontal, or are inclined at the same angle in the same direction, they are said to be 'conformable,' *i.e.*, they were deposited in regular sequence, without any interval or break between. There are a few exceptions to this rule, where beds not really conformable appear to be so, but this is owing to accidental coincidence of dip, and may here be disregarded. When one set of rocks reposes on another with a different dip, they are 'unconformable,' the upper set resting, not on the surface, but on the denuded edges of the lower. A similar appearance to unconformity may be produced by an 'overlap' in strata which are really not unconformable, by the thinning out of one bed, or of more than one bed, in the series.

Frequently beds are fractured, those on one side of the crack, usually termed a 'fault,' having been displaced vertically, by upheaval or by letting down, in regard to those on the other side. The displacement is called the 'throw,' and may be a few inches only, a few feet, or even several hundred yards. Beds which are faulted to any extent, of course are not continuous, neither are the lower rocks where there is unconformity, and this failure in continuity exercises a great influence, beneficial or otherwise as the case may be, upon the extent of mineral deposits, and especially upon the flow of underground waters.

Mining operations.—It has been shown that the 'Minerals' and 'Metals' for which mining is carried on depend entirely upon the nature of the rocks of which they form a part or with which they are associated. Upon this depends also in a great measure the kind and the cost of preliminary borings, sometimes incurred without the slightest prospect of success, of the main shafts or pits, when the deposits sought for have been discovered, and of the actual mining operations; but these are much more influenced by the relation of the rocks in and beneath which these are performed.

'In no respect do collieries differ more from each other than in the quantities of water which they encounter, either in the mining or in the subsequent working of their mineral. In one case a retentive clay cover may prevent the access of surface water which in another may pass in abundance through a sandy or a gravel alluvium. In certain districts water-bearing measures of an almost fluid consistency must be passed through, whilst in others the comparatively tight coal measures may at once be entered. Frequently the strata above and below the coal are so compact as to render the workings actually too dusty and dry; but instances are common enough in which water makes its way through the roof stone, or through the coal itself, and adds difficulties and expense to the whole of the operations. When the measures through which the pit is sunk consist of stony rock they are often allowed to stand open, but when shales preponderate it has to be walled with brick or stone, to which in some cases, as against the influx of water, wood or cast iron may be preferred. But when the measures are covered by other and more absorbent strata, saturated with water, the winning of a colliery becomes a most serious undertaking, tasking the energies of the best men, and sometimes collapsing after a ruinous outlay. Examples of these difficulties are afforded by surface beds of sand and gravel, and by the well-known red sand under the Magnesian Limestone. One of the most serious questions to be solved by the coal-viewer in the very outset is the system by which he means to work his mineral; and in order to form a judgment upon this head it is important that he should not only be acquainted with the various modes in use elsewhere, but should have acquired a knowledge of the peculiarities of the seams in his own district. Where the beds have a definite dip in one direction, the working pits are usually placed as far towards the deep as it is convenient to go, so that underground the coal may be brought down hill to the pit-bottom. Should the strata lie in a trough, the pits may advantageously be placed in its middle line, so as to command the coal on both sides.'*

Cuttings.—It is evident from a consideration of the relation of two or more rocks to each other, as defined above, that it may affect engineering works even to a greater degree than the nature of the rocks on which such works are situated. In nothing is it more evident than in railway works that the relation of two beds—to take a simple case—differing from each other in kind, but having, it may be, the same dip, will be such as to increase the cost of any work upon them, if risk to its stability is to be avoided. Let us suppose a railway-cutting of moderate depth to traverse two beds of different character, one a water-bearing sand resting evenly upon a tenacious clay, both dipping, transversely to the line, at an angle of 3 degrees.

* Smyth's 'Coal and Coal Mining,' 2nd edition, pp. 109, 111, 121, 122, 175 (Lockwood : 1872).

The slope on the higher side of the cutting will, in a very short time, be scored by a series of weeping springs along the line of junction, which will surely, although perhaps slowly, cause serious slips unless means be taken for their prevention. Should the dip lie in the same direction as the fall of the ground

Fig. 1.—Railway cutting in pervious and impervious strata. *a*. Pervious stratum. *b*. Impervious stratum.

—which is, however, unusual—the flow of water will probably be greater at some times than at others, the springs will, perhaps, be intermittent. Any structures, such as bridges over the cutting, must then have their footings well down into the clay, not on it, for its surface, even many feet from the ground-level and under an equal thickness of solid-looking rock, would be absolutely unsafe as a foundation.

If in the above simple instance of the effect of the relative positions of strata (neither of which by itself would demand special care) precautionary measures be necessary, much more must the frequently intricate relations of the rocks receive careful consideration. The beds may dip much more rapidly, the water-bearing strata may be numerous, and the geological structure may otherwise be complicated by faults or by local unconformity. (See also p. 42.)

Tunnels.—A previous knowledge of the relations of strata to each other is still more desirable in tunnelling operations; indeed, it would be impossible to insist too strongly on the necessity for all geological details being known in regard to a hill to be pierced in that manner. Not only might the style of working be varied, but the form or strength of the tunnel itself might perhaps be altered with advantage, to suit either the varying pressure or the peculiarities of rocks known to occur in the centre, different to those exposed in the cuttings at either end. Even the gradients might require to be modified according to the existence or non-existence of faults or of springs in the body of the hill, which if not previously detected would be discovered when it is too late to make any alteration.

If the surface of a hill be carefully examined and the boundary lines of the beds of which it is composed be accurately surveyed, by the methods described in Part II., their dip can then be ascertained, and the lines of all faults laid down with a reasonable approach to accuracy. The position of all the rocks within the hill can then be defined as well as at the surface, consequently the points at which they will be met with in the tunnel can be accurately determined. Even in cases where the all-important point of the amount of dip cannot be obtained from actual sections, it may be worked out—by a method hereafter explained—from the boundaries, or other definite lines, if these have been laid down on the plan with precision.

In a series of beds passed through by a tunnel, those which are uppermost generally occur near the centre of the hill, unless they dip in one direction only, when the upper bed of all is, of course, at one end. This is owing, when it occurs, not merely to the fact of the gradient rising from each end towards the interior, but to a well-known geological phenomenon. As a general rule beds are found to dip from each side into a hill or ridge, beneath which they occupy a synclinal trough; the statement being, however, limited to hills and ridges as such, and not

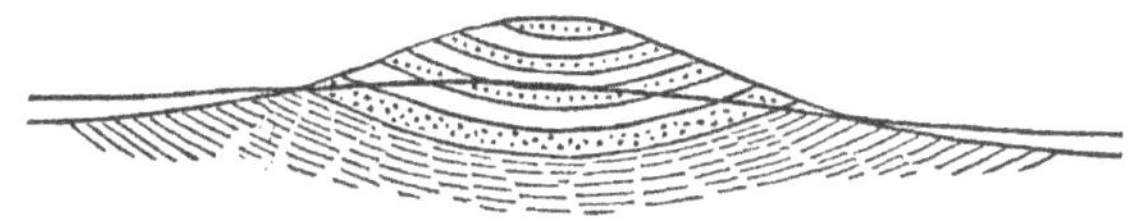

Fig. 2.—Railway Tunnel through a ridge, the beds of which occupy a synclinal hollow. The thick black line represents a longitudinal section of the tunnel.

to include escarpments. Therefore, as the tunnel proceeds, beds are successively pierced which are higher and higher in the series, until the uppermost of all that are met with is reached somewhere beneath the axis of the elevation.

Banks.—At first sight the relation of the rocks beneath the surface may not seem to have any direct bearing upon railway embankments, and similar artificial accumulations of material. But there are ways in which it does now and then greatly affect the cost of such works, and, what is equally important, that of bridges and culverts erected beneath them.

A line of railway does not usually run in the direction of dip of the strata, but rather at right angles to it, consequently nearer to that of the strike of the beds, as it follows, in a general way, the contour of the country. The dip of the rocks beneath, whether it be great or small, therefore, is generally away from the railway, either to the right hand or to the left, as the case may be; but there will be distances for which the line of dip may more or less coincide with that of the railway. In such cases, if the dip be of any appreciable amount, say exceeding five degrees, the bank, when it attains to any height, will be very likely to force the beds beneath it over each other along the planes of bedding. This gives rise to slips, sometimes of great extent, and involving much loss of material; they are usually sudden and liable to repetition, causing great expenditure for pile-driving, timbering, and other preventive measures. But if the liability to slip, owing to the dip of the beds—which usually are, in such cases, alternating clays and dissimilar deposits —be previously entertained, it may be minimised by 'running the tip ahead,' and working backwards with the bulk of the material.

If the dip be into the hill whence the material comes, there is no risk of slips happening from this cause, but it is occasionally the other way, when they are sure to occur. Such slips, if quite forward, are sometimes of slight consequence only, as the bank is then advancing in the right direction. But should any bridge or culvert have been built in their path, ready for backing up, the consequences may be serious, for such structures are then almost sure to be overthrown, unless the bulk of the work be done in the backward manner mentioned above.

Reservoirs and Canals.—The preceding remarks apply equally to the construction of water reservoirs and canals, but, in regard to them, it may be added that the geological structure of a country must greatly affect the supply of water to canal-feeders and its retention in natural reservoirs. The 'head' of water that can be maintained in such reservoirs, generally formed by a dam across some minor valley and supplemented by pumping, is limited by the springs which may occur within its area; not by the amount they are capable of yielding, but through other phenomena described in Part III., which may render useless, to a great extent, the large expenditure frequently incurred for pumping. The attempt is, in fact, often made to obtain a head of water which the geological conditions, left to themselves, make absolutely impossible; therefore, if these be understood, and measures taken accordingly, much cost and useless trouble may in many instances be saved.

Main Drainage.—A smaller, but not unimportant matter, is the difficulty sometimes experienced in carrying out main-drainage works, owing to the surface and other springs tapped by the excavations. Several instances have occurred where, owing to the quantity of water thus met with, the plans, after commencement of the works, have had to be altered, at great disadvantage; others, in which the pipes have been imperfectly laid, or have afterwards settled in the quicksands, so that the joints have ever after been unsound. In consequence, they have admitted the spring waters, and thus added, perhaps, several hundred pounds a year to the cost of pumping, besides deteriorating the value of the sewage for irrigation or precipitation. These results have happened, in some cases beyond hope of remedy, not because the geological details, including a knowledge of the springs, could not have been ascertained in time, but simply because they have been ignored. The remedy for such a state of things, where one is practicable, is to be sought in the plan described in Part III., but it must always be adopted within certain limitations, and, indeed, should only be allowed under official supervision.

Foundations.—One observation may here be made with regard to foundations, whether of dock works, bridges, or buildings. From the preceding notes it will have been seen that however solid the stratum may apparently be in which the excavations are made for foundations, the calculations as to stability are incomplete and liable to error—as the works are to unforeseen catastrophe—unless all the relations as well as the nature of the rocks beneath have been ascertained and taken into consideration.

Water Supply.—The question of supplying the inhabitants of towns and villages with good drinking water forms one of the chief problems of our day. It is essentially a question for engineers, but a knowledge of geology is indispensable to him who would attempt its satisfactory solution. An immediate improvement in the water supply of any district, and in its sanitary conditions generally, is one of the practical results that may reasonably be expected to arise from the working out of its geological structure—that is, from a knowledge of its rocks and of their relation to each other. For the available supply of water in any given locality is not by any means proportionate to its rainfall; as by the widely spread water-bearing beds it is to a great extent equalised. The supply to be obtained by boring down to deep-seated springs is practically inexhaustible, being scarcely, if at all, affected by drought, and these springs form the only source on which can be placed a full reliance.

The phenomena of springs, and of the sources of supply to Artesian wells, both of which are practically important, are entirely dependent on stratigraphical and physical features. An explanation of them has been given in Part III., as well as of the reason why sometimes salt waters occur far inland, and fresh-water springs beneath the sea; why some waters are chemically pure, whilst others are saturated with mineral salts. And it may be added that the localities are few beneath which such springs may not be found at a greater or lesser depth, according to their physical and geological features.

Dampness.—Another point affected by exactly similar conditions to those that govern water-supply is the dampness of a locality; a condition which sometimes renders almost uninhabitable what would otherwise be a desirable and healthy situation. It may arise either from the physical conditions of elevation, situation, rainfall, and so on, or from those of a purely geological nature. Frequently it is owing to the saturated condition of the water-bearing beds immediately beneath, and probably close to, the surface; in other words, by the proximity to the ground level of the general water-line of the district. This is a question that should influence the choice of situation for all public buildings, and, indeed, for private houses also, where the circumstances are such as to admit of selection. It is fully considered in Part III., a section being devoted to the subject of 'Sites,' as one which merits, but seldom receives, much attention.

Disease.—A knowledge of the undoubted relation existing between subsoil and disease must also be beneficial, especially to those who may be seeking for themselves a new home. This question cannot be fully entered on here, but the results of certain official inquiries into one branch of the subject are appended. In the 'Report of the Medical Officer of the Privy Council,' for 1867, pp. 14—17, and 57—110, is discussed, from several points of view, the interesting question of the connection between the geological structure and the consumption death-rate of a district. After careful consideration of all the facts and statistics adduced, the following suggestive and valuable conclusions were arrived at, and may be considered as fairly well established:

(*a.*) That on pervious soils there is less consumption than on impervious soils.

(*b.*) That on high-lying pervious soils there is less consumption than on low-lying pervious soils.

(*c.*) That on sloping impervious soils there is less consumption than on flat impervious soils.

(*d.*) These inferences must be put along with the other fact, that artificial removal of subsoil water, alone, of various sanitary works, has largely decreased consumption. From which follows the general inference, that wetness of soil is a great cause of consumption.

If this one disease can be so influenced that its ravages in a district may be, as they have been, lessened one half by a simple draining of the land, it may reasonably be assumed, that the power of other diseases also is more or less dependent on certain physical conditions, which are susceptible of natural or artificial modification.

PART II.

PROCEDURE IN THE FIELD.

CHAPTER I.

METHODS EMPLOYED IN GEOLOGICAL SURVEYING. MAPS AND SECTIONS.

Methods employed in geological surveying—*Determination of rocks*—*Tables.*

Methods employed in Geological Surveying.—Geology, as a science, relates the history not only of the rocks of which the earth's crust is composed, that follow each other in regular if somewhat disturbed sequence, and form the indestructible pages of a record on which that history is imprinted. It describes also the successive races of animals and plants that have flourished on its surface or in the ocean, and whose remains, now petrified, were enclosed in the rocks at the time of their formation. An acquaintance with the organic remains enclosed, as fossils in the strata, is therefore of great scientific value in determining the geological age or position of the rocks, but it is to a great extent unnecessary for practical works. What is required for mechanical operations and for sanitary purposes is a knowledge of the stratigraphical geology of a district; that is to say, of the nature of the local rocks and of their relation to each other.

If we would make a series of drawings to show the geological structure of any locality, we must first trace upon a map the boundaries of its rocks, and thus define the area that each formation occupies. We must next ascertain the angle at which the rocks dip away from the surface, and then, aided by our notes, construct a section, which shall portray their underground extension and relative position. We must further ascertain the kinds of rock of which the beds consist, by their general appearance and the aid of simple tests in the field; or if necessary, by others, more complicated, applied to detached specimens at home. Thus, to form an accurate idea of the rocky structure of a district, and to represent and describe its geological features, three operations have to be performed:—

(1). The character and peculiarities of the strata which crop out at the surface determined.

(2). The boundaries of the different rocks laid down upon a map.

(3). The dip, if any, and the underground continuation of the beds worked out and shown upon a section.

The methods of geological surveying adopted in the carrying out of these operations are described in the following order:—*Determination of rocks; Geological maps; Geological sections.*

Determination of Rocks.—It is obvious that for engineering purposes a great deal depends upon the physical characters of a rock, whether regarded as a building material in itself, as a source whence such material is derived by any process of manufacture, or as a substance possessing any influence upon construction, stability of work, or any other practical considerations. Therefore it is highly necessary for the geological engineer to be able not only to determine the class to which any particular specimen belongs, but also to form some reliable ideas as to what substances enter into its composition. There are simple tests, for application in the field, to ascertain the class of rock under examination; and more delicate tests, by which the field results may be checked and extended. In most cases the directions given below will go far towards accurate determination; but it will be advisable to consult works devoted to the subject, in the examination of difficult specimens.

To ascertain the kind of rock exposed in a pit or quarry, a fragment is detached from that part which has been least altered by the action of the weather. A good-sized piece of the rock is broken off, and afterwards reduced by chipping into a square lump; good edges are thus obtained for observation of its texture by the aid of a pocket-lens or magnifi

In the field, the first test may be made for hardness—a character determined with reasonable accuracy by the facility or difficulty with which the specimen can be scratched, if at all, by a pocket-knife. The result will show to which of the three main divisions of the following table the rock may be referred—scratched with ease, with difficulty, or not at all. It may then be tested for effervescence by dilute hydrochloric acid, a small bottle of which is always carried for that purpose in geological surveying. The fact of its effervescing rapidly, slowly, or not at all, will narrow still more the limits of the list in which it is included. Further determination is then made by the additional tests of texture, colour, etc., given in the second column of the table, which in most cases will settle at least the class to which a mineral or a rock belongs. The texture is observed at the chipped angle of the specimen; it may be crystalline, glassy, compact, earthy, granular, or laminated. The peculiarities of fracture, lustre, and streak are valuable aids in a precise examination, but are omitted from the table for the sake of greater simplicity, as without them results may be obtained sufficiently near for all practical purposes.

The behaviour before the blow-pipe of many substances is given, to distinguish some which cannot otherwise be separated. This instrument is of great assistance in discovering the exact character of rocks and ores; a microscope also is invaluable for the same purpose. There are some rock substances which may be chemically analysed without much trouble; but works, like those named below, specially devoted to blow-pipe, microscopic, and chemical analysis should be consulted for the methods of determination of the ultimate constituents of a specimen.

The Study of Rocks. Rutley (Longmans).
Introduction to the use of the Mouth Blowpipe. Scheerer and Blanford (Williams and Norgate).
Chemical Geology. Bischoff.

TABLES FOR APPROXIMATE DETERMINATION OF MINERALS, ORES, AND ROCKS.

FIELD TESTS.	ADDITIONAL TESTS.
1. *Those which are easily scratched by a knife,* and—	*Texture, usual colour, behaviour before the blow-pipe, etc.*
(*a*) *Effervesce rapidly,* are—	
Calcite (carbonate of lime)	Crystalline, white or tinted, infusible, reduces to quick-lime
Satin spar (carbonate of lime)	Fibrous „ „
Marble (carbonate of lime, with some silica, alumina, etc.)	Crystalline, various colours
Limestone „ „	Compact, sometimes oolitic „
Chalk „ „	Earthy, white or pale yellow
Galena, lead ore (sulphide of lead)	Crystalline, dark grey colour, decrepitates and fuses, yielding metallic globule.
Chalybite, iron ore (carbonate of iron)	Crystalline, brownish colour, blackens and becomes magnetic
Clay ironstone „ (impure „)	Concretionary, brown „ „
(*b*) *Effervesce slowly.*	
Dolomite (carbonate of lime and magnesia)	Crystalline, white or tinted, infusible, reduces to quick-lime
Magnesian limestone („ „ with some silica, alumina, etc.)	Compact or granular, greyish colour
(*c*) *Do not effervesce.*	
Selenite (sulphate of lime)	Crystalline, white, exfoliates, becomes opaque
Gypsum	Compact or minutely crystalline, white or tinted „ „

Fluor spar (fluoride of calcium)	Crystalline, white or purple-tinted, fuses to a clear bead, which on cooling becomes opaque
Sandstone, micaceous	Granular or laminated (laminæ glistening), various colours
Hornblende schist	Foliated, black
Glauconite (silicate of iron and potash)	Compact, olive green, fuses to magnetic glass
Chlorite (silicate of magnesia, alumina, and iron)	Compact, foliated or granular, dark olive green, fuses on thin edges with difficulty
Chlorite schist (chlorite, quartz, etc.)	Foliated, green
Mica (silicate of alumina, etc.)	Various colours, plates elastic, fuses on thin edges only
Mica schist (mica, quartz, etc.)	White or green (folia hard and glistening), dark grey
Talc (bisilicate of magnesia)	White or green, plates not elastic, fuses on thin edges with difficulty
Steatite, 'soapstone'	Compact, whitish, fuses with difficulty
Serpentine rock (silicate of magnesia)	Compact, dark olive-green, fuses on thin edges
Fuller's earth (silicate of alumina)	Earthy, greenish-brown, fuses to porous slag
Slate „ „	Compact (cleaved), dark grey
Barytes (sulphate of baryta)	Crystalline, white or tinted, decrepitates and fuses with difficulty
Blende, zinc ore (sulphide of zinc)	Crystalline, black or brown, infusible alone
Blue vitriol (sulphate of copper)	Crystalline, with soda yields copper bead
Copper glance, copper ore (sulphide of copper)	Crystalline, lead grey, fuses, yielding copper bead

Copper pyrites, copper ore (sulphide of copper and iron)	Compact, brass-yellow, fuses, yielding magnetic bead
Copperas (sulphate of iron)	Compact, green, fuses with borax yielding a green glass
Rock salt	Crystalline, tinted, decrepitates and fuses
Graphite, black lead (carbon)	Crystalline, foliated, black, infusible
Coal (impure ,,)	Compact, black

2. *Those which are with difficulty scratched by a knife,*

and—

(*a*) *Effervesce rapidly.*

Calamine, zinc ore (carbonate of zinc)	Compact, greyish, infusible alone, fuses easily in borax.

(*b*) *Effervesce slowly.*

Limestone, siliceous	Compact, various colours

(*c*) *Do not effervesce.*

Felspar, orthoclase (silicate of alumina and potash)	Crystalline, sometimes compact, white or pink ; fuses on thin edges only
Felspar, oligoclase (silicate of alumina and soda)	,, ,, white or tinted, fuses with difficulty to clear glass
Felspar, common	Compact ,, ,,
Hornblende (silicate of lime, magnesia, etc.)	Crystalline, green, brown, or black, fuses easily to magnetic globule
Hornblende rock	Compact, green, brown, or black, fuses easily
Augite (silicate of lime, magnesia, etc.)	Crystalline, greenish-black, fuses easily to grey glass
Hypersthene (silicate of magnesia, and iron)	Crystalline, brownish-green, fuses easily to black enamel

Hypersthene rock, 'greenstone' (oligoclase and hypersthene)	Crystalline, greenish-black, fuses easily
Felsite (orthoclase and quartz)	Compact, grey, weathers white
Dolerite (oligoclase and augite)	Crystalline, granular, dark grey
Basalt	Compact, black
Diorite, 'greenstone' (oligoclase and hornblende)	,, green, weathers brown
Porphyrite (oligoclase and hornblende, etc.)	,, pink, brown
Gabbro, 'greenstone' (oligoclase and diallage	Crystalline, greenish
Trachyte	Compact (feels rough), grey
Obsidian	Glassy, brown or grey
Phonolite, 'clinkstone'	Compact, grey, weathers white
Apatite (phosphate of lime)	,, infusible
'Coprolite' ,,	Concretionary, brown
Specular iron, iron ore (peroxide of iron)	Crystalline, steel grey, infusible alone
Hæmatite ,, ,, ,,	Reniform, red
Limonite ,, 'brown hæmatite'	Compact, brown

3. *Those which cannot be scratched by a knife,*

and—

Do not effervesce.

Granite (quartz, felspar, and mica)	Crystalline
Gneiss (,, ,, ,, in folia)	,,
Syenite (,, felspar and hornblende)	,, more durable than granite
Quartz (silica)	,, white or tinted. infusible alone
Quartzite (fine grains of quartz in siliceous matrix)	Compact, granular
Quartz sandstone (,, ,, if a slow effervescence the matrix is calcareous)	,, ,,

Flint (silica, not quite pure)	Compact, black or grey
Hornstone	,, ,,
Chert	,, various colours
Iron pyrites (bisulphide of iron)	Crystalline, bronze-yellow, fuses to globule attracted by magnet

CHAPTER II.

METHODS EMPLOYED IN GEOLOGICAL SURVEYING (*continued*).

Construction of geological maps—*Geological surveying*—Example—Dip.

Geological Maps.—It is very essential when tracing and mapping geological boundaries to have a good map of the district to be thus surveyed, for, however accurately those lines may be laid down, any errors on the map must affect the after calculations. All the physical features of the district should be distinctly indicated on the map, such as hills, rivers, and streams, and heights above the sea level in figures here and there are of great advantage. Maps drawn to a scale of 1 inch to a mile answer well for general purposes; if great accuracy be required, those on a 6-inch scale should be used. Both kinds are issued by the Ordnance Survey; of the whole country on the smaller scale, but on the larger scale those of a part only have up to the present time been published.

Contour lines are engraved on the 6-inch maps, and running, as they do, through all the points where a horizontal plane at any given height would intersect the surface of the ground, are of great assistance in geological surveying. To the eye accustomed to them these lines convey at a glance the physical geography, or the actual shape of a tract of country, its hills and valleys, its precipices and ravines. Contours, of course, run in a V-like shape up the valleys, in straight lines on even flanks and ridges, and sweep in curves round the outlines of the hills; their variations are numerous as those of the features themselves, but these kinds of form prevail in all.

The three following propositions afford considerable aid to those engaged in geological surveying:

(*a*.) The boundary-lines of horizontal strata exactly coincide with the contours.
(*b*.) The boundary-lines of strata dipping towards a hill are less winding than the contours.
(*c*.) The boundary-lines of strata dipping from a hill are more winding than the contours.

Therefore:

(*a*.) One point through which the boundary passes being ascertained, the line will thence follow the contour exactly, so long as the beds continue horizontal.
(*b*.) In this case every point on the line of strike, at the same level as one through which the boundary passes, must be also on the line of boundary. A line somewhat less curved than the contour, being flattened in proportion to the dip, represents the line required.
(*c*.) When the strata dip with the slope, the line must be discovered at several points, and these united by exaggeration of the contour, but with the windings reversed when the dip exceeds the slope of the ground.

Geological Surveying.—Useful as the above propositions are, the ground must nevertheless be gone over, and the actual line followed, for dip may change where alteration is least expected. Faults also may occur; these will suddenly terminate a boundary and introduce a fresh line, representing the broken ends of the strata that have been upheaved on one side of it, or thrown down on the other.

A geological map is one which, as we have seen, defines the area occupied by the surface of each formation, or by its denuded edge where it comes to the level of the ground. It follows that to construct such a map with accuracy every part of the surface within any area to be geologically surveyed must be examined. If by any means the nature of the rocks over a given area be proved, say by borings or trial-holes, one in every acre of ground, and the varying results be shown by different colours, a geological map would be roughly presented. But it would be an approximation only, for the lines of division between the rocks would still remain to be shown.

Similar evidence to that afforded by trial-holes may be obtained, in many other ways and with much less trouble, as to what is the uppermost stratum at any given point, or any number of points; and for our present purpose it is in regard to this stratum only that the information is required. Road-side banks and ditches, even of moderate depth, will all yield evidence as to the kind of rock at the particular spot where examined. It is obtained by picking into the bank, or the side of the ditch, and by cutting at the face of each exposed section; but in all cases the actual stratum beneath the soil and rain-wash must be determined. In the absence of all such sections, the surface-soil may be turned aside, here and there, along the probable line of boundary. The heaps of different stuff thrown out from their holes by moles and rabbits will afford indications, and search in ploughed fields will generally lead to the discovery of lumps of the rock being traced. Lastly, the eye, by practice, will enable us to judge from the soil itself what is the rock beneath, from which it has been formed by a process of disintegration. Another important point to be noted is the existence of springs, as these indicate not only a change from pervious to impervious strata, but also the actual lines of division. All the results of the examination of the ground must be entered on the map by some mark or symbol, in the exact situation of each observation, and from these the geological lines will be drawn.

Where a rock comes to the surface, the area it occupies is bounded by two lines: one is called its 'line of outcrop,' and coincides with its upper edge; the other coincides with its lower edge, and is called its 'line of boundary.' The former is of course the line along which it crops out from beneath a higher stratum, the latter marks its lower and outer margin or boundary, and is the same as the line of outcrop of the rock immediately beneath. Sometimes the geological will coincide with the topographical lines, but it is not often the case, as the intricate windings of the natural divisions of the rocks adhere, more or less closely, as previously explained, to the contour lines of the country.

Traversing.—In walking over a district for the purpose of mapping its strata, it is well to follow some regular plan. The lines of ditch and fence may be taken up and down alternately; if too far apart, they can be left, and the ground between walked over in search of evidence. By thus following one fence, for, say, half a mile in any direction, and returning by another the width of one field apart from the first, and by repeating the process at the distance of another field, all details concerning a considerable area may be collected. This plan may involve the drawing of several geological lines as the boundaries of the strata that occur are crossed and recrossed. It is frequently found more convenient to commence and follow out one line only at a time; this is done by walking along a ditch or fence until the line is discovered, crossing the field, and following the next fence until the line is again crossed, and so on to its termination.

Example of Geological Surveying.—The nature of some of the rocks occurring in the district to be surveyed having been previously determined, their relation will be worked out, and the area occupied by each defined by drawing its boundary-line in the following manner:—Let it be assumed, for sake of illustration, that the evidence obtained, whether by trial-holes or by any of the preceding methods, has established the existence of limestones over all the higher ground, of clay upon the flanks of the hills, of ferruginous sandstone in places along their foot, with clay again in the plain beyond.

Starting from some point on the table-land, and following a line of fence, the surveyor notices that the soil is light, and full of fragments of a whitish limestone, varying in size from small grains to lumps as large as a hen's egg; and in the ditch he occasionally sees the beds of limestone in their undisturbed position. Just below the point where the ground begins to decline towards the valley he observes, in a slipped portion of the bank, rock of a more sandy nature, full of concretions of a yellow, rusty-looking substance, which turns out to be a workable ironstone. Near this point is a small spring, the water from which has worn away the ground into a hollow and deep channel, exposing the blue clay just below. It is evident that the pervious limestones overlie the impervious clay, by which the water that has percolated through them has been thrown out on the hill-side; also that there is between them, a bed of ironstone of slight thickness. The line of division, of course, passes through this spring, the exact position of which must be fixed on the map, by compass-bearing or otherwise. A short line is then drawn in pencil from this point to either hand, in the same direction and of the same form as a contour-line would have at that elevation.

This line of boundary will again be met with in returning by the next fence, although it may not again be exposed by a slip or indicated by a spring; but when the surveyor arrives near the spot through which he expects it to run he seeks for it by some of the methods already described. He may not be able actually to see the junction of the beds, but he will presently be at a spot where limestones occur just above him and blue clay below, and here, of course, must be the boundary. At this spot is, perhaps, a small plantation, a bend in the fence, or some other mark shown on the map. He will draw contour-lines from this point also to either side, one of which lines will be found to unite with that drawn from the first fence in this direction. It is for the present assumed that the beds are horizontal: should it prove otherwise, the pencil lines, if curved, must be flattened or exaggerated, accordingly as the dip is into or away from the hill.

But to return to the line of fence along which the mapping was supposed to be commenced. After passing the spring, the surveyor walks for some distance over a clay of bluish colour, which is seen every few yards in the ditch that runs straight down the hill. The soil is hard, and in many places cracked; this would sufficiently indicate the nature of the stratum beneath, even were it not visible as it is in the sides of the ditch. Towards the lower part of the slope the soil gets lighter, being covered by a sandy wash from the hill above, and presently a pond is met with, which shows, when its margin is picked into, signs of another bed of sandy ironstone. This is noted on the map by a symbol in its exact position, and left for the present; it may be of trifling local occurrence, or it may be part of an important deposit. Beyond this is blue clay again, covered by a foot or two of sandy wash, but exposed wherever a ditch, pond, or other excavation through the surface-soil lays it open to inspection.

Returning by the second fence in the direction of the high ground, an exactly similar, but reversed, sequence of strata is observed. There is, however, a brick-pit situated just at the foot of the slope, in the vertical face of which is seen a bed of calcareous and ferruginous rock, two feet thick, with blue clay above and below. This bed is evidently continuous, judging by contour, with that noticed in the pond, therefore a line is drawn to connect the two points, and following the shape of the ground. Continuing his walk up the hill, the surveyor again traverses the blue clay until he reaches the small plantation above mentioned; beyond this point he meets with limestones only, and in these there are several quarries on the table-land.

The sequence of beds in the area covered by this short walk is thus established, as:—

Limestones, in horizontal beds.
Ironstone, thickness not proved.
Blue clay.
Bed of Ironstone, 2 feet.
Blue clay.

And in this order of succession they would be met with in a boring commenced on the higher ground. The thickness of each would be determined without a boring, if a section of the surface were taken, to show the position of their outcrop, for the beds are, so far as we have seen, horizontal, and present no signs of unconformity.

In this way simple boundary-lines are discovered and drawn; they follow the form of the ground, which, indeed, is mainly due to the varying hardness and changing dip of the rocks where they rise to the surface. For, in accordance with those characters, the denuding agencies have worn away the rocks; that is to say, the minor features of a district are hills or hollows, according to the power its rocks possess of resisting denudation. As dissimilar rocks must thus make a change of feature along their junction, a knowledge of the fact is, as we have seen, turned to account, and is of great service, in drawing their lines of boundary.

CHAPTER III.

METHODS EMPLOYED IN GEOLOGICAL SURVEYING—*continued.*

Dip and Strike—Rules for finding dip—*Example.*

Dip and Strike.—The term 'Dip' has frequently been used, and it is scarcely necessary to explain that it means the angle at which the bedding planes of strata are inclined to the horizon. A vertical plane intersecting the bed where it makes the greatest angle with the horizon, is in the 'direction of the dip.' Another line, at right angles to this, must necessarily coincide with that part of the rock which is horizontal, and would be seen in such a position in the face of a quarry cut back in the dip's direction. This second line is called the 'strike' of the rock, and if the surface of the ground were a level plane the outcrop of a stratum would always coincide with the strike, and be, throughout its entire length, at right angles to the dip. But as the surface is uneven, the outcrop winds accordingly, crossing and recrossing the general line of strike, which must perforce remain the same until the dip assumes an altered direction. And as the dip varies, so does the width of outcrop of beds or formations of a given thickness.

Clinometers are generally used for measuring the

angle of dip, but other instruments answer the purpose equally well if they indicate the amount of inclination. It is most frequently in mines, quarries, brickyards and similar places, that sections of the rocks are seen, but it should be remembered that these exposures may not be in the true dip's direction, and that the apparent dip only can thus be taken. This must be observed in two or more faces which make a considerable angle with each other, the lines of their direction being laid down on paper; the result may then be worked out by the rule given below.*

Note.—The true dip may be greater, but it cannot be less, than that seen in any face of a section open to observation.

Rule.—When two observed dips incline from or towards the angle enclosed by their lines—A B, A C, in fig. 3, p. 75—the true dip is at right angles to a line thus laid down:—Set off from the angle on each one of the two lines of apparent dip, a number of units corresponding to the number of degrees of dip observed along the *other* line. A line connecting the two points coincides with the strike, and is consequently at right angles to the true dip's direction. Instead of the angle of dip, the proportionate incline may be set off on *its own* line, the result obtained, of course, being also a proportionate incline.

When one observed dip inclines from and the other towards the angle—as A D, A C, fig. 3—the true dip can be worked out by prolonging one of the lines of apparent dip beyond the point of convergence; both

* Published by the author in the *Geological Magazine*, May, 1876.

apparent dips will then run either from or towards the angle thus formed, and the true dip can be found by the above rule.

Should the amount of dip be considerable, and great accuracy be required, set off the tangent of the angle of apparent dip. For, except in the smaller angles, there is an error arising from the difference between their circular measure and their tangent, but in most cases this is too slight to be of any practical value.

When the direction of a dip has been thus worked out, it becomes necessary to ascertain its amount, also, from the apparent dips observed in the quarry. From any two observed dips the amount and direction of the true dip may be obtained by calculation, but for all practical purposes the results arrived at in the following manner are sufficiently accurate and much more expeditious :—

Construct on one of the observed lines a right-angled triangle representing the apparent dip, its base being equal to the length set off along that line. The perpendicular is equal to the amount which the bed falls in a distance equal to the length of the base of the triangle. The bed falls the same amount in the shorter length at right angles from the line of strike (previously drawn) to the angle of the quarry. Another triangle constructed with this shorter base and the same perpendicular gives the exact measure of the dip required.

The true dip and its direction may also be worked out, and the existence of two different dips discovered in any locality, where no sections of the beds are exposed. This is done by first accurately surveying some definite

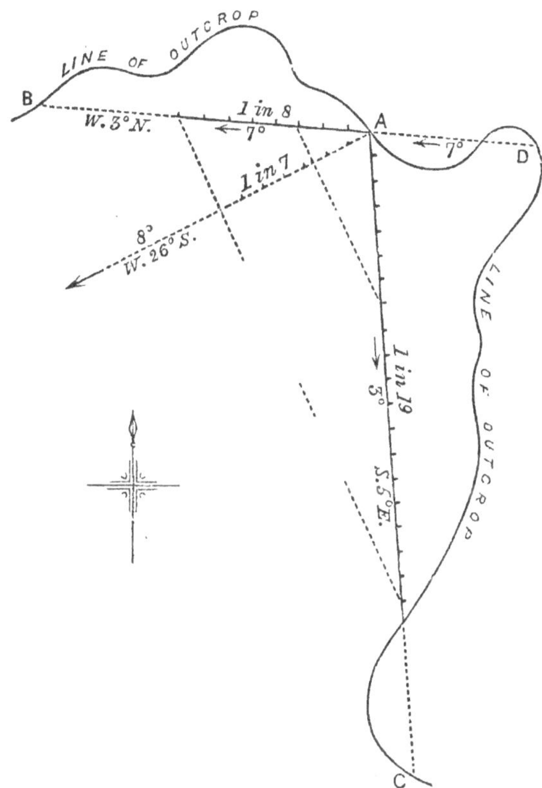

Fig. 3.—Diagram to illustrate the Rule for working out the true dip, from observed apparent dips—as angles or proportionate inclines—and from three points on an outcrop.

A B, 75 feet in 200 yards = Incline 1 in 8.

A C, 63 „ „ 400 „ = „ 1 in 19.

line, such as a thin band of rock, or a line of division between two beds, and by then taking the relative heights of three of the most suitable spots upon it for that purpose. These spots should be so chosen that lines connecting them will form a near approach to a right angle; the distances apart must be either measured or scaled from the map; the heights can be taken by a level, or by an aneroid if extreme accuracy be not required. The points thus selected will afford the same information as if the stratum along the lines connecting the points of observation were exposed in two faces of a quarry. The lines represent the direction, and the differences in height will give, with the lengths, the proportionate incline or, by aid of a diagram, the amounts of the two apparent dips, from which the amount and direction of the true dip can be found by either of the methods described.

Example.—The following is an example of working out dip from three points of outcrop by means of proportionate incline :—

Three points, A, B, C, are on the line of outcrop of a definite stratum; A is 200 yards (measured W. 3° N.) from, and 75 feet above, B; 400 yards (measured S. 5° E.) from, and 63 feet above, C.

The fall from A to B is thus found
to be at the rate of . . . 1 in 8
The fall from A to C at the rate of 1 in 19

Lay down these lines on paper, as in fig. 3, and
Set off on A, B 8 units of length.
" " A, C 19 " "

Connect the twopoints thus set off, by a line that will be in the direction of the strike, and at right angles to that of the true dip, which proves to be W. 26° S.

The amount is thus ascertained :—

The bed falls at the rate of 1 in 8 along A, B—1 foot (or other unit) in the 8 set off along that line (and similarly 1 in the 19 along A, C)—it falls an equal amount in the shorter length of a perpendicular let fall from the angle at A to the line of strike already drawn: that length scales 7 units, giving a proportionate incline of 1 in 7 (equal to 8°), in the case in question.

As dip may vary rapidly, this is not always a strictly accurate method, but with care it will give approximate results in all cases, and is especially valuable in the surveying of districts for the estimation of their water-supply.

For a description of the instruments used in geological surveying and for tables of dip, proportionate inclines, etc., the reader is referred to the treatise previously mentioned—'Field Geology'—which is intended more especially for geological students, and in which the subject is treated at greater length than is considered necessary in a work designed for those to whom such instruments are necessarily fa

CHAPTER IV.

METHODS EMPLOYED IN GEOLOGICAL SURVEYING—*continued.*

Example of surveying faulted area—Drift deposits.

Example of Geological Surveying.—A somewhat more complicated example of geological surveying may now be given, involving the application of the methods described of determining the nature of rocks, of observing dips, and of tracing irregular boundaries, faults, and unconformities. The surveyor usually obtains his evidence for the lines and draws them at the same time; in addition to this, he also, in the same traverse, notes all exposed sections of the rocks, measures all dips, determines the nature of the various beds, and collects all other geological information.

The area to be surveyed is a mile wide from west to east, and a mile and a half in length from south to north; its main physical features are a ridge of high ground on the south, sloping generally down towards the north, with a stream running in an easterly direction across a depression in the centre. It is shown as a completed geological map in fig. 4, p. 80, from which, however, the roads and other topographical details have been omitted, in order that the geological lines may be more easily followed in the description of the surveying operations.

Commencing near the north-east corner and walking in a westerly direction, it soon becomes plain to the surveyor, from the nature of the soil, that a sand, or sandstone, forms the stratum beneath. There are no ditches, a fact which also indicates subsoil of a light description, and one that requires no draining; the absence of these, however, prevents, for a time, any other than surface indications being obtained. But a quarry, No. 1, is presently met with, and its exact position is laid down on the map as ascertained by compass-bearing. One observation is taken upon some distant object, perhaps a church; another upon a second object, the lines crossing each other at an angle, as nearly as may be, of 90°.

The face of the quarry No. 1 shows the following beds in descending order, which are thus entered in the note-book:—

Brown and reddish-coloured freestones in beds 1 ft. to 2 ft. thick, with partings of sandy clay	10 feet
Red marl, with lumps of a white and pink crystalline substance	2 ,,
Red freestone, as above	4 ,,
Red marl, not bottomed	3 ,,

The apparent dips are to N.W. 3° 20′, and N.N.E. 4° 20′.

The dip, worked out, by the rule given in p. 73, from the apparent dips observed along two sides of the quarry, proves to be due north 5°, this is entered

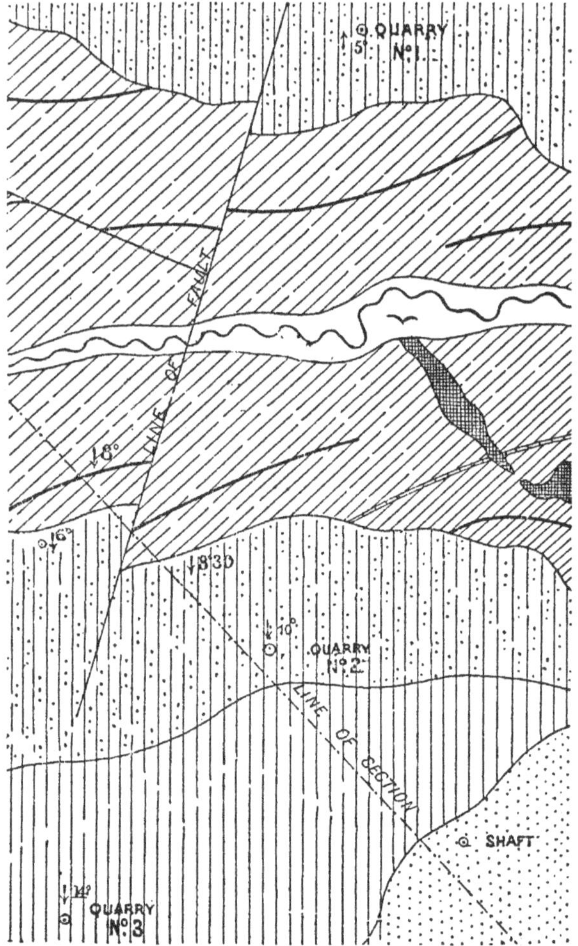

Fig. 4.—Map of Area geologically surveyed.

Reference to Fig. 4.

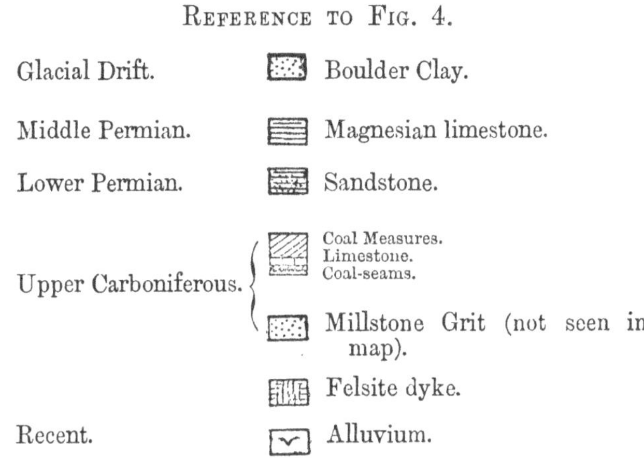

on the map in its right position, the amount in figures, with an arrow pointing in the dip's direction. This direction is transverse to that of the general shape of the ground, consequently the feature, or contour-line —modified as explained on p. 64—will be found to correspond, or nearly so, with the strike of the beds, or possibly with the boundary of the formation to which they belong.

The freestones are easily scratched by a knife, and do not effervesce upon application of the dilute acid; they have a very fine granular texture, with glistening particles in rather indistinct lines of lamination. From these characters the beds are concluded to be micaceous sandstone; the partings of shale are calcareous, being slightly effervescent. The pinkish substance included in the red marl is equally soft, and does not effervesce; it is minutely crystalline, as seen under the lens, and proves to be gypsum. (Table, p. 58).

Continuing to advance in a westerly direction from the quarry, the surveyor finds that near the edge of the map the sandy soil disappears, and is replaced by one which is darker in colour and more clay-like in character. A little to the south many lumps of coal are lying about, and in a ditch an actual seam of coal, whence they were probably derived, is seen in place, with shales above and below. The outcrop of this seam is drawn, and the boundary of the sand at the same time, by traversing the ground, so as to cross and recross their lines. These are found to gradually approach each other, so that the coal-seam—or rather its outcrop—dies away at the boundary of the sand; the coal, with its associated shales, passing beneath the red sandstones.

A little further east the edge of the sand suddenly appears to trend nearly southward, running indeed almost at right angles to its direction as far as this point. The exact position of the line here is not very distinct, so it is drawn somewhat doubtfully, when at about five chains to the south it is found to resume its original direction. There is evidently some break in the continuity of the line here, and it will have to be ascertained whether this is due to a fault, as is probably the case, judging from the apparent shift in outcrop.

From this point the sandstone line nearly follows the contour, almost to the edge of the map, where another coal-seam abuts against and passes beneath its boundary. These lines of outcrop of the coal-seams, abutting as they do against the winding boundary of the sandstone, testify to a difference in the line of strike, and a consequent difference in the direction of dip of the two formations. It will be at once apparent, without an actual section showing the beds in their relative position having been met with, that the fact observed is a positive proof of the existence of either a fault here also or an unconformity. The second coal-crop is mapped in another walk to the westward, and is found to be broken in a manner similar to that of the sandstone boundary. Another breakage occurs in it further on; however, the outcrop is mapped, for the present, as nearly as may be from the evidence obtainable, and the surveyor proceeds to the other side of the stream, having first drawn a line corresponding to the northern margin of the alluvium.

In following the alluvium line on its south side he meets with a wall-like mass of rock standing well up above the surface of the shales which form a flat in the bottom of the valley, and cutting across their strike in a south-easterly direction. A detached specimen of this rock can be scratched only with difficulty, and it does not effervesce with acid, therefore it is a silicate; being grey in colour, weathered white on its outer coating (which effervesces along its inner margin), and compact in texture, it is almost certainly Felsite, the mass being either eruptive or intrusive. It proves, when afterwards mapped, by the change it makes in the shape of the ground, to be not interbedded with the coal and shales, but an eruptive dyke breaking through the strata without the slightest conformity to their planes of stratification.

The next easterly traverse, after the completion of the southern edge of the alluvium, crosses another coal-seam, and the edge of some red sands and marls similar to those on the north side of the stream. A small section in the shales associated with the coal shows the dip to be S. 8°; the sands where exposed in a deep ditch are seen to be dipping also in a southerly direction, at an angle of 6° only—an additional proof of the unconformity between them. In a small pit further east the dip of the sands has increased to 8° 30′; and it must be observed that this is in an exactly opposite direction to the dip obtained in quarry No. 1, showing that the beds form an anticlinal arch, its axis coinciding with that of the little valley. But in the neighbourhood of these sections a break occurs in both the coal-crop and the sandstone boundary, similar to the breaks on the other side of the stream; and it is seen, when the lines are mapped, that another nearly straight line would pass, with one exception, through all the points where the boundaries and outcrops are broken. This fact testifies to the strata being dislocated by a 'fault,' of which the sudden change in dip previously observed is also an indication. A line drawn through all the fractured ends of the beds represents the line of the fault; on its western side the beds have been thrown down to an extent that will afterwards be ascertained. A second fault, nearly at right angles to the first, accounts for the break in one of the coal-seams somewhat to the westward of the other fractured portions.

The boundary of the southern portion of the sandstone is laid down in this traverse, with two outcrops of coal-seams and one of a thin band of limestone, all following the physical feature, or shape of the ground.

The limestone is found to be cut across by the felsite dyke, of which the mapping is completed by carefully following its irregular edge, the line being drawn as it is discovered, for it has no relation whatever to the general contour. (See page 83.)

In the next walk to the westward, another set of beds is found coming on above the red sands and marls, noticeable at first by the great change in the colour of the soil, and by lumps of limestone scattered over the higher portion. A section will perhaps be found in these new beds; but in the meantime the boundary can be readily traced by ditches, the differences in soil, and similar surface indications. About half-way across the map another quarry is met with in the sandstone, and the exposed section is thus noted :—

QUARRY No. 2.

Alternating and varying beds of micaceous sandstone and sand, indistinctly bedded, of reddish colour, with shaly divisions and patches of gypsum .	30 feet

Dip S. 10°.

The uppermost bed in this section is a rubbly mass of calcareous shale and weathered pieces of limestone, probably remnants of a deposit once overlying the sandstones, but removed by denudation.

The face of the quarry on the south shows the beds in a horizontal position; the dip observed on one side is therefore correct, in amount and direction.

In returning eastward, and not far from the corner of the map, a large quarry is met with in the beds mentioned as probably coming on above the red sandstones and marls. It is worked for building purposes, the material being of excellent quality, and, although easily worked when first quarried, it is compact and durable: the stone can be easily scratched by a knife, it effervesces slowly with acid, is compact and of grey colour, and proves to be Magnesian Limestone. The face of the quarry is very irregular, but a projecting angle at one part gives an opportunity of taking two apparent dips with their direction. These and their result, with the notes on the beds, are given below:—

QUARRY No. 3.

Magnesian limestone, tabular and jointed, in beds from two to four feet thick, with thin earthy layers between	40 feet
The upper beds are light-grey in colour, with many fossils and some concretions. Lower down the stone gets harder, and in some cases granular. The lowest beds are of darker colour, more gritty texture, and, in some parts, distinctly crystalline.	

Apparent dip to S.S.E, 13° } =S. 14°.
" " W.S.W., 5° }

Judging from the observed dips, and the position of Quarry No. 2 below that of No. 3, the red sandstones noted in Quarry No. 2, and similarly in No. 1, must pass in under the Magnesian Limestones, between it and the Coal Measures. They probably are therefore the sandstones and marls of Lower Permian, as the limestones are of Middle Permian age.

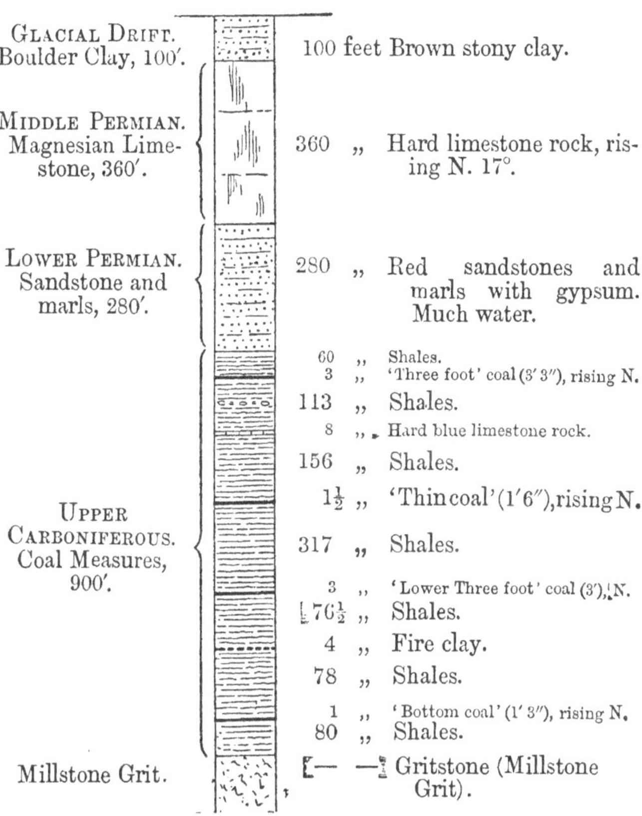

Fig. 5.—Section of Beds passed through in sinking the Upcast shaft at .... Colliery.

The Magnesian Limestones are found to be covered, at the south-east corner of the map, by a brown sandy clay, unstratified and full of fragments of rock of various kinds, and—in this respect like the felsite dyke—its boundary fails to follow the shape of the ground. This clay is known as 'Boulder Clay,' and is one of the members of the 'Glacial Drift,' the boundaries of all which deposits, whether clays, sands, or gravels, are most irregular, as explained on p. 89. Being closely followed, this irregular deposit is found to fall rapidly away to the north, and evidently must soon mask the boundary of the limestone, perhaps that of the sandstone also, even if it does not descend quite into the valley.

In walking over the Boulder Clay in search of sections, and to complete his traverse of the area, the surveyor comes to the pit's mouth of a colliery. The coal-seams down to and through which the shaft has been sunk, and in which mining is carried on, are no doubt those which rise to the surface in the lower ground to the north, and were found to be dipping in a southerly direction. Figure 5, p. 87, represents a vertical section, made from particulars kept at the manager's office, of the strata passed through in sinking the shaft.

Drift deposits.—The different sets of fossils enclosed in the stratified formations are remains of plants and animals which flourished respectively during the periods of their deposition. These remains—some sets of which are of a tropical, some of a temperate, others of an arctic type—prove many important changes in the climate prevailing in the same area at different periods of the earth's history. During at least a part of the time included in the Glacial period an arctic climate prevailed over Britain, covering the land with ice and snow. Of this there is ample evidence, also of concomitant physical conditions, by which the geological products of that age are strangely affected. There were then great oscillations in the relative level of land and sea, so that, in turn, every part of the surface of these islands was more than once subjected to the abrading action of moving ice. This agent denuded the old surface in a rough and irregular manner, leaving equally rough and irregular masses of the abraded material.

The deposits of this period are chiefly unstratified clays, in which are seen numerous rounded and angular lumps of older rocks, laminated and sometimes contorted brick-earths, huge moraines of angular rock fragments, unstratified gravels, and false-bedded sands. All these occur in a partial and, in one sense, an uncertain manner, as regards thickness, extent and distribution, the only regularity being their succession (on a large scale) in definite sequence. It matters not, for practical purposes, what form of ice has produced these results, whether land-ice, coast-ice, or icebergs, or whether one only or each of these agents of glacial action may have had its share in the production of the deposits which are, however, admitted by all to have had a glacial origin.* As the clays and gravels occur in the irregular manner described, their boundaries will not have the same mathematical relation to the

* The views of the author on this subject are expressed in a Survey Memoir on 'The Geology of Cambridgeshire.'

shape of the ground as those of stratified beds. In this respect they are analgous to those of the eruptive rocks (p. 84), and are traced in a similar manner by being closely followed. Evidence of the lines must be obtained at frequent intervals, without trusting to contour or stratigraphical continuity, if any scientific or practical conclusions are to be drawn from the extent of the beds or their mode of occurrence.

The climate has, since the Glacial period, gradually ameliorated, and as the land arose from its last submergence (a fact also generally conceded) spreads of gravelly material were formed, here and there, upon the surface, derived from the waste of the glacial deposits or of older rocks as the case may be. These, now reduced to mere remnants, are found usually on the highest lands and in patches of no great extent.

Next in order are some sands and gravels, also occupying high lands as well as ridges within valleys, which indicate the earlier positions of the rivers. In some instances the old courses may have been coincident with those which the rivers now occupy at a lower level, but were usually distant from them; sometimes parallel with, but more frequently transverse to, the lines of the modern streams. At lower and lower elevations, are found other terraces of river gravel, each marking stages in the excavation of the valleys in which they occur, until those bordering the rivers are met with at a level just above them, which belong to the most recent stage of all before the rivers scooped out their present channels.

As a rule, these old gravels, which contain prehistoric human relics and the remains of extinct mammalia, are not continuous, but occur in lines of small and usually elongated patches, capping mounds and ridges, and more rarely banked against the flanks of existing hills. It is evident that they were not deposited in such positions, so far, at least, as the mounds and ridges are concerned, but that the line of ground they cover was a valley at the time of their deposition. The patches, therefore, rest in old hollows or depressions, which give to their lower layers a synclinal form that has, in a great measure, contributed to their preservation, while the surrounding ground, formerly higher than that which they occupy, has been removed by denuding agencies. This trough-like form of section in these gravels has a bearing upon their water-yielding capacity; all the water within them, at a lower level than their boundary, is retained, but it would otherwise be thrown out along the margin, and it forms the source of supply to the surface-springs upon which many of the smaller towns and villages depend. ('Field Geology,' p. 312.)

The lines of these deposits are easily drawn; they follow the contour, or nearly so, from any point where the boundary may be discovered, except where the beds rest against the flank of a hill, and do not form a ridge or mound of their own. In these cases allowance must be made for the bed thinning out against the slope of the hill, and in drawing all boundaries of sand or gravel the area enclosed by the lines should not spread as far outwards as the surface-soil would indicate its extension. The line must be kept up, or back, to allow, according to the probability in each instance, for the material which will have been washed down the sloping ground and thus gives a deceptive appearance of the bed to some distance below its actual boundary.

CHAPTER V.

METHODS EMPLOYED IN GEOLOGICAL SURVEYING (*continued*).

Geological sections—*Practical value—Levels—Filling in.*

Geological Sections.—The amount of information furnished by a geological map is greatly increased if the map be illustrated by a section showing the underground extension of the rocks occurring within the area portrayed. When a geological map has been properly made, it conveys a valuable but still limited knowledge of the sequence of the beds, and if the works for which such knowledge is required be affected only by their surface characteristics a section is unnecessary. But for all purposes in which the more deep-seated rocks and their relation to each other, and to those near the surface, would exercise an influence, a section is invaluable. This may be either a simple diagram or a section drawn to scale: for the latter a line of levels is of course requisite, showing with accuracy the form of the surface along the route to be thus illustrated. The geological details may be filled in either from surface evidence and details of actual exposures or borings along the line, or from surface evidence, physical features, and scientific deductions.

Diagrams are useful, whether exaggerations or not,

to illustrate any particular points, as by their aid the mind is enabled to grasp details much more readily than from verbal description alone. Sections drawn entirely from observed facts are of course the most trustworthy, but it rarely happens that any number of exposures, or borings of any depth, can be obtained in the line of country which the section is intended to traverse. Therefore geological sections are generally made from evidence obtained at the surface, all the available data being studiously collected. The dips are worked out from these, all the known natural phenomena bearing upon the hidden rocks of the district are carefully considered, and the probabilities of other and more deep-seated influences upon them are taken into consideration. From the results thus obtained, supplemented and checked, perhaps, by an occasional well or boring, good geological sections can usually be drawn. It may be confidently asserted that if due attention has been paid to the points indicated, such sections may be relied on as presenting, if not an accurate representation, certainly an approach to the facts as they occur in nature, sufficiently near even for all practical operations.

Practical Value of Geological Sections.—The knowledge afforded by geological sections of the position of the various rocks a long way beneath the surface of the earth is (as urged on pp. 44—47) especially valuable to the engineer in laying out his lines for tunnels and deep cuttings. It may often be desirable to follow the course of a particular bed, or series of beds, on account of its value as building material or ballast, or of the facility with which it can be penetrated, or because it would form, owing to its strength or continuity, a good roof for tunnelling operations. On the other hand, it may be equally desirable to avoid certain rocks, on account of their bad qualities in one or all of these respects, or because of their liability to slips, or from the probability of their yielding a troublesome flow of water; but unless the internal structure of the hill be known, the desired end is almost certain not to be attained.

In designing such works to the best advantage, a point would be selected for the commencement of the tunnel or cutting where the most suitable beds come to the surface at the proper level for the construction of the line, as determined by the equalisation of earthwork. Then the strike (not necessarily the local outcrop) must, if possible, be followed, a course which would insure the continuance of the work through those beds so long as the base of the cutting or tunnel should remain at the same level. Some amount of fall, in one direction or in both, would, however, generally be necessary, and by swerving a little to one side of the line of strike—that is, to the rise of the beds, according to the amount of dip and requisite fall—the same result of continuance in the same beds would be insured. Not only would the work be of similar nature throughout, but also the resulting material, whilst the risks involved in a change of strata in tunnelling would be avoided—indeed, the desirability of thus selecting a line is so great that it is scarcely possible to over-estimate its importance.

Of equal value is such knowledge in the estimation of water-supply from deep-seated springs, for upon the relative position and permeability of the rocks far beneath the surface depends the success or failure of the projected operations. The artesian-well engineer who can work out these problems is able to select the most promising position for a boring, to calculate approximately the depth to which it must be carried, and to estimate the quantity and quality of the supply to be obtained. He can also make his preparations beforehand for passing through those springs which are to be rejected, and he knows when and where all expenditure in boring would be useless for the purpose in view. In Part III. this subject is more fully discussed, and an example is given of the methods of procedure and calculation.

Levels.—It is necessary to have the surface of the ground along a line of geological section accurately represented, otherwise errors will creep in where perhaps they are least expected. The means by which a line can be drawn to represent the surface are sufficiently numerous, and need not here be described; by some approximately, by others every rise and fall, however slight, is shown with the greatest accuracy. There are the precise methods adopted for engineering works, in which the level and theodolite are used; and others, less accurate, which are however very serviceable on account of their quickness or the portability of the instruments employed. These are frequently of much use in unravelling the intricacies of a difficult piece of geological mapping as the work proceeds, and the necessary observations can be taken by the surveyor while engaged in traversing the country for that purpose. Sections are thus drawn from contour maps, from the Ordnance 'heights above the sea,' or from observations taken with an aneroid barometer, care being taken to eliminate the error arising from change of atmospheric pressure during the time of making the observations.*

All sections, by whatever mode the data are obtained, require that heights be known at certain definite points, varying in distance from each other according to circumstances, the intervening portions being sketched in to correspond with the surface configuration. For diagrammatic sections the distances apart of these points can be scaled from the map, but for accurate work measurements by the chain are required.

It is usual in geological sections to refer all heights to the Ordnance datum, or 'level of the sea,' which is the level of mean tide at Liverpool. In the sections published by the Government Geological Survey a datum is assumed 1000 feet below the level of the sea. All the points on the Ordnance maps, where heights are figured, are bench-marks, the levels of each having been ascertained with reference to the Ordnance datum.

As a general rule, the rocks will have been mapped before the levels are taken for a section to illustrate their mutual relations, therefore the section line can be laid out to correspond, as nearly as may be, with the dip's direction. But sometimes it is found convenient to construct a geological section along a line which

* The methods usually adopted are described, with examples of levelling by aneroid barometer, in 'Field Geology' (Baillière), 1879, p. 128.

does not run in the direction of the dip, when the beds must be shown inclined at a smaller angle in proportion to such divergence, the difference being found by diagram or by calculation.

Filling in Sections.—For filling in the geological details to sections, the points of boundary and outcrop crossed by them are either scaled off from the map or plotted from notes of their position in the measurements taken in running the surface levels. The beds are drawn inwards, from these points, with the proper inclination along the line of section, true or apparent dip as the case may be, and with all the faults,

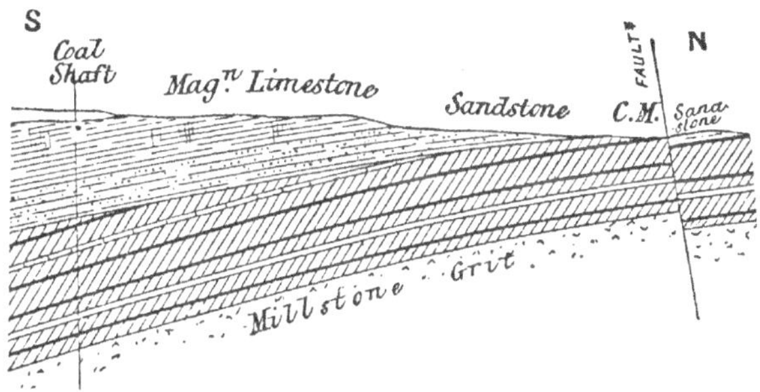

Fig. 6.—Section showing the underground extension of the rocks which come to the surface in the area surveyed, as represented in the geological map, Fig. 4, p. 80.

flexures, and contortions that can be ascertained; dotted lines should be used where these details are uncertain.

The dip of all the beds in the section, fig. 6, is necessarily shown much less than it would be if the line of the section coincided with that of the dip, from which it deviates 45°, for the purpose of crossing the fault and of utilising the evidence afforded by the shaft of the coal-mine. The dip of the limestone, for instance, is at its boundary (calculated from the observations in the quarries) about 10° 30′; it is drawn 7° 30′ only, as the result of its 45° variation from the true dip's direction. The Coal measures are dipping at 22°, and are shown less in proportion. This difference in amount of dip, taken in actual section through the two formations, proves the unconformity inferred from the gradual approach of their lines of boundary. (See page 82). The amount of 'downthrow' of the fault is calculated from the known dips and the lateral shift of the beds; that is, by finding from these the difference in depth below the surface of the same bed on either side of the fault where crossed by the line of section.

PART III.

ECONOMICS—MATERIALS, MINERALS, AND METALS; SPRINGS AND WATER SUPPLY.

CHAPTER I.

MATERIALS, MINERALS, AND METALS.

Economic products of the Recent and Tertiary Rocks.

THE following brief notes on materials, minerals, and metals are given as affording suggestive indications of what may reasonably be expected to occur within the limits of the different formations distinguished by colours and symbols on geological maps. Those who are engaged in designing or executing engineering works are recommended to ascertain from good maps of this kind what formations occur within the area with which they are immediately concerned, and to trace out for themselves, according to the directions given at page 54 *et seq.*: the local nature and exact extent of the deposits. These vary in character and economical value too much in different areas for any general description to be otherwise of any great practical utility.

Beyond a doubt there are many beds of good material, or of other commercial value, and many prolific mineral veins, the existence of which has hitherto escaped notice. Others have been merely indicated on maps and in works primarily prepared for scientific durposes; and their extent or value can only be ascertained by strictly local and detailed investigation. For example, there are numerous bands of ironstone and of phosphatic nodules, much too thin or disconnected to make an appreciable or 'mappable' outcrop, which consequently do not appear even on the official geological maps, and which receive only casual mention in the scientific memoirs.

The following table shows the general sequence of the geological formations, and the nature of the prevailing rocks in each; their economic characteristics being more fully described under the headings of the group to which they belong.

TABLE OF STRATA.

RECENT.

Alluvium, *Silt*, *Peat*, Blown Sand, Shingle.
Valley *Gravels, sand, brick-earth.*
Terrace *Gravels* (*Palæolithic implements*).
Plateau *Gravels and loam*

GLACIAL.

Boulder Clay. Till. Moraines.
Sands, Gravels and loam.
Brick-earth.
CROMER, and elsewhere.—Submerged Forest Beds.

TERTIARY.

EAST ANGLIA.—Crags. *Shelly gravels.*
DEVONSHIRE.—Bovey Beds. *Lignite and clays.*
I. OF WIGHT.—Fluvio-marine series. *Clays, marls, sandstones and limestones.*

London Basin.—Bagshot Beds. London Clay. Lower London Tertiaries. *Sands, pebble beds, clays and variegated loams.*

Cretaceous.

Upper Chalk, with flints. Lower Chalk, mostly without flints. *Chalk, sandy limestones, flints.*

Upper Greensand. Gault. Lower Greensand. *Sand, sandstones, firestones, limestones, clays and 'coprolites.'*

Weald of Kent and Sussex.—Wealden Beds. *Clays, sands, sandstones, limestones and ironstone.*

Oolitic.

Purbeck and Portland Beds. Kimeridge Clay. Coral Rag. Oxford Clay. Corn-brash. Great Oolite. Inferior Oolite. *Limestones, clays, sandstones, grits, flags and ironstone.*

Liassic.

Upper Lias, Middle Lias or Marlstone, Lower Lias, Rhætic Beds. *Clays, limestones and ironstone.*

Triassic.

Keüper, *marls, sandstones, gypsum and rock-salt.*

Bunter, *sandstones and pebble-beds.*

Permian.

Magnesian *Limestones, sandstones and marls.*

Carboniferous.

Upper Carboniferous. *Coal, shales, sandstones, grits and iron stone.*

Lower Carboniferous. *Limestones, sandstones, clays, Trappean rocks, ores of iron, lead, zinc,* etc.

Devonian and Old Red Sandstone.

Sandstones, slates, marls and iron-ore. Granitic and Trappean rocks.

SILURIAN.

Sandstones, grits, flags, shales and limestones.

CAMBRIAN.

Slates, sandstones, flags, Granitic and Trappean rocks, ores of copper, lead, zinc, etc.

LAURENTIAN.

Granitic and metamorphic rocks, sandstones, etc.

Great care is necessary in the selection of building materials, especially of those required for important public works, where strength and durability are indispensable qualifications. In the selection of a stone, it is not merely in the testing it by hardness, composition, and appearance, that judgment should be displayed, but also in ascertaining the conditions under which it lay before removal from its position in the quarry. For instance, many of the oolitic limestones, which are hard and compact, appear to the eye as excellent building stones, but split up into thin plates and fragments on exposure to the weather, from the effects of which they have been hitherto preserved by several feet of superincumbent clay.

A method, adopted by M. Brard, of testing the relative value of various stones as building material, in regard to durability, especially applicable to oolitic and other calcareous rocks, is to be found at page 464 of the second edition of Ansted's 'Geology.' The action of frost is induced by the crystallisation of Glauber's salts in a saturated solution of which cubes of the stone are boiled; they are then suspended over the solution, and plunged into it for the removal of efflorescent crystal, during a period of four days. The particles scaled off by the process are then weighed, and form a comparative index of the durability of the stones under examination.

Another point to be borne in mind is that some finely laminated sandstones and shelly limestones having a similar structure will weather off in flakes if the stone be used as flags for paving, or be placed with its lines of laminæ vertical in walls, or parallel with any external face of a structure. With their edges only exposed to the weather these stones last as long as any other, and in this way are frequently used as coping stones to rough walling. Again, 'Sandstones being composed of siliceous grains, they are more or less durable according to the nature of the cementing substance, while limestones are durable in proportion rather to the extent in which they are crystalline' (*Ansted*). But the best and most reliable test of all, so far as durability is concerned, is for the engineer or architect to examine for himself, old structures known to have been built from certain definite beds taken from quarries in the neighbourhood.

Professor Hull, in his treatise 'On Building and Ornamental Stones,' says that 'the density of a building material is to a certain extent a test of its compactness and durability.' Also that 'in selecting a stone special attention should be paid to the climate of the locality in which the building is to be erected. The chemical constituents of rocks are of great importance in reference to their durability under certain circumstances of the atmosphere, and conditions of climate. The presence of smoke, of sulphurous, hydrochloric, and other acids powerfully aids the destructive effect of rain or moisture; limestones and dolomites are especially subject to disintegration from the influence of rain charged with acid. The best kinds of building stone for smoky and wet climates are siliceous sandstones, formed of grains of quartz, cemented together by a siliceous or felspathic paste.'

The stone which is most suitable for building purposes is rendered, by the same properties, the best also for ballasting lines of railway; that which falls to pieces by the action of the weather must, of course, be scrupulously avoided. The stone employed for road metal should be tough as well as hard and durable, and, if possessing an uneven texture, it is to be preferred on that account. For this purpose 'sandstone is better than limestone, and hard limestone better than slate; while basalts and granites are exceedingly good or exceedingly bad, according to the proportion of alkaline earths which they contain' (*Ansted*).

In the following notes upon the occurrence of building materials amongst the various geological formations, the rocks of igneous origin are mentioned in connection with the products of the sedimentary deposits with which they are associated, their geological age being unimportant for practical purposes.

Recent Deposits.

Alluvium and River Gravels.—The materials furnished by the alluvial deposits of existing rivers and streams are comparatively few and unimportant. Consisting mainly of muddy loam and silt, the alluvium makes excellent pasture land, but yields no building material beyond an occasional bed of brick-earth, sometimes worked to a depth of a few feet in the absence of more suitable deposits. The river silt, when coarse, sharp, and clean, like the 'above-bridge sand' of the Thames, is valuable for building purposes, especially for stucco and similar work; when fine in grain, this sand is useful also, in the process of brick-making.

Deposits of brick-earth occur here and there at a level slightly above that of the present rivers, and are sometimes of excellent quality, as well as of considerable extent and thickness; such are the beds extensively worked on the banks of the Thames below London. The gravels of this series are generally fine, the stones of which they are made up being usually small and subangular; the material they afford is used for road-mending and ballasting railways. A greater part of the metropolis stands on gravel of this age and description. Bath-bricks are manufactured from a recent sand full of siliceous remains of minute organisms, and terra-cotta has been made from an alluvial clay in Devonshire. There are other deposits of gravel similar to the last, but usually coarser, which occur as terraces, at higher levels, and mark the former course of existing or of ancient rivers. These gravels are sometimes remarkable for their enclosing rude flint implements, relics, so far as can at present be positively asserted, of the earliest inhabitants of these islands. (See page 90.)

Glacial Deposits.

It was mentioned in Part II. that the surface of many parts of the country is covered by drift deposits, and that these occur alike on hills and in valleys, in a very irregular manner. As a rule they have no relation to the present form of the ground, but mask the older rocks over some large areas, and are at the same time totally absent from others. The accumulations of this period consist of clays, gravels, sands, and brick-earths, varying in composition and appearance, but all containing fragments of the older rocks.

The clays are occasionally used for brick-making, it being essential, however, that they are worked up in a pug-mill to get rid of the included fragments; in some instances they are so calcareous as to be burnt for lime. The chalky boulder clay has been much used in former times for marling land, and with excellent results. The brick-earths form extensive beds, and are largely worked in the Eastern counties; from one of these deposits the famed Woolpit bricks are manufactured. There are many large but irregular spreads of glacial gravels and sands, varying in coarseness, the former yielding good material for making roads and ballasting railways, the latter for building and brick-making. Some of the gravels are made up more or less of rounded pebbles, and afford useful stones for paving and ornamental building purposes.

The gravels and sands of this and the preceding series make light soils which require high farming, their yield being also greatly dependent upon the wetness and dryness of the seasons. The brick-earths form loamy soils, especially valuable for hop-growing and for market-gardening; the boulder clay makes excellent corn land, its fertility and ease of working being greatly increased by early ploughing and exposure to the frosts of winter.

The recent and glacial deposits occur under conditions favourable to the production of surface-springs. (See p. 133.) But the beds are so irregular, and sometimes thin out so rapidly, that their occurrence at any point cannot be depended on without careful examination.

TERTIARY ROCKS.

The *Pliocene* deposits are almost entirely confined to the Eastern counties, and yield but little material calling for notice here. They consist in great measure of shelly sands or 'crag,' used for gravelling garden walks and private roads, with occasional beds of clay and brick-earth. One notable product, however, of these deposits is the Suffolk bone-bed, a seam of 'coprolites,' or phosphatic concretionary nodules, containing 45 to 60 per cent. of phosphate of lime, which has been largely worked as a source of artificial manure, but is now almost exhausted.

The *Miocene* beds are rare in this country; they include the 'Bovey coal,' seams of lignite worked near Bovey Tracey in Devonshire, and some beds of true coal in the islands off the west coast of Scotland. Some felspathic clays belonging to the same series, and of considerable thickness, occur in the same district as the Bovey lignites; they are used for pottery, and yield pipe-clay of good quality.

The *Eocene* deposits are more extensive and important than those of the preceding divisions of the Tertiary period, and occupy the areas known as the London and Hampshire basins. They consist of clays, loams, marls, and sands, with occasional beds of rolled flint pebbles, and more rarely of fuller's-earth. Pipe-clay occurs in some of the beds, and seams of lignite are not uncommon. The clays and loams are much worked for brickmaking, especially the mottled loam of the Woolwich and Reading series, and the well-known London clay. The latter contains layers of septaria, or clayey limestones, called cement-stones, which are dredged off Sheerness, Harwich, and elsewhere, for the making of Roman cement. In places there occurs a calcareous sandstone; ferruginous sandstone and the many-coloured Alum Bay sands are found in the Isle of Wight, with a remarkably pure and white sand employed in the manufacture of glass.

The Eocene soils are various as the beds themselves, the Thanet and Woolwich sands forming light land, intermediate in character between the rich loams of the Reading beds and the barren sands of Bagshot-heath and of the same age elsewhere. The London clay is tenacious, and forms a stiff soil, valuable as pasture, and as arable fairly productive when well-drained; the marls of the Hampshire basin are of a fertile character.

Where the Tertiary sands form the surface stratum, they yield water, but its quality—like that from all cultivated lands—should be, in every instance, examined. The Bagshot sands throw out a line of springs around the lower portion of their boundary, by which it can be readily followed. The beds below the London clay formerly contributed greatly to the supply of wells in and near London, but their yield has been much reduced by constant pumping, and at the present time is quite insufficient; they form, however, a useful source of supply, where a very large quantity is not required.

CHAPTER II.

MATERIALS, MINERALS, AND METALS—*continued.*

Economic products of the Secondary Rocks.

UPPER CRETACEOUS.

THE *Chalk* is an earthy limestone, consisting of about ninety-five parts in a hundred of carbonate of lime; generally white and soft, but containing some hard, sandy, grey beds, which are quarried for building purposes. Some of these harder beds, although easily cut, are well able to resist the action of the weather; they have been largely employed for tracery in church windows, and during the lapse of several centuries have suffered very little from atmospheric influences.

A bed of hard grey sandy chalk, slightly yellow in colour, which might well be called a fine-grained calcareous sandstone, occurs at the top of the Chalk-marl; it is known as the 'Totternhoe Stone,' and has been much used in building. The white chalk is burnt into lime, the grey chalk into an excellent hydraulic lime; the softer and more pure beds are levigated and made into whiting.

The Upper Chalk contains many horizontal layers of flints, useful for building and road-making; in some parts where building material is scarce, the walls of houses are built almost entirely of flints, having brick or free-stone at the angles. In a few old church towers the necessity for even this addition was dispensed with by their being built circular on plan. The flints are used also in the manufacture of china and flint glass. At the base of the Chalk occurs a bed which in places abounds in phosphatic nodules; it has been extensively and very profitably worked in many localities on and near its outcrop in Cambridgeshire, Buckinghamshire, and elsewhere. The mineral phosphate, of which these nodules contains about 55 per cent., is converted by chemical treatment into biphosphate of lime, and extensively used as artificial manure.

The Lower Chalk and the Chalk-marl make fertile soils, much more so than that of the lighter Upper Chalk, which on its higher and more exposed portions, such as the North and South Downs, is scarcely covered by vegetable mould. It, however, produces a short but constant growth of herbage, on which sheep are profitably fed, and in some seasons the cultivated portions will yield fair crops of turnips.

The Chalk is an excellent water-bearing formation, not only on account of the quantity it will yield under certain conditions, but also of the quality of its water. Springs once met with in the Chalk are constant if not uniform in their yield, owing to its extent, to its great thickness of rock of fairly equable character, and to its absorbent power. Chalk absorbs water more rapidly than any other solid rock, which, however, passes very slowly through its mass; in consequence of this peculiarity the large springs are met with only in the numerous fissures (forming natural collecting galleries) by which some parts of the formation are traversed in every direction.

Although the Chalk is, at or near its outcrop, a reliable source of water supply, it does not do to depend on meeting with powerful springs in it, beneath the Tertiary strata, except where the water occurs within it under certain conditions—where fissures are proved by existing wells to be numerous, or where the boring is to be carried down below the line of saturation. (See page 138.) It is an unusual circumstance, but borings have been carried down even several hundred feet in the Chalk without tapping a single spring of any consequence.

The *Upper Greensand* consists mainly of sands, and sandstones of varying degrees of hardness, frequently argillaceous, sometimes calcareous, sometimes micaceous; the calcareous sandstones are quarried for building, and occasionally burnt for lime; the more siliceous beds also furnish excellent building stone. This formation contains many beds of firestone, notably in Surrey; the malm-rock, which occurs in Hampshire; many beds of chert, which makes good road metal; also phosphatic nodules in sufficient quantity to be of economic value within the Wealden area.

Exposed in many valleys cut through the chalk, the Upper Greensand gives rise to a particularly fertile soil, yielding good returns for the higher systems of agriculture. Where it occurs in sufficient thickness, it yields water of good quality, which is held up by the impervious Gault clay beneath.

The *Gault* is a persistent formation of stiff blue clay, calcareous, with occasional bands of septaria, and having, near its base, workable beds of phosphatic nodules; being well suited to the purpose, it is dug in many places for brickmaking. At the base of the Chalk in Norfolk and Lincolnshire occurs the Hunstanton limestone; neither the Upper Greensand nor Gault being present, this rock is probably the representative of one or both of those formations. Although good for pasture-land, the Gault soils otherwise are not generally productive.

Lower Cretaceous.

The *Lower Greensand* formation varies considerably, comprising in different localities sandstones, sand, limestones and clays. The sandstones are mostly ferruginous, sometimes occurring as coarse grits quarried for road-mending, and are known locally by the name of 'carstones.' The sandy beds in the Weald are termed 'hassock,' and alternate with beds of Kentish rag, a cherty limestone of bluish-grey colour, sometimes burnt for lime, and largely employed in the construction of important buildings. The clays include some calcareous bands, seams of phosphatic nodules, and layers of septaria used for making Roman cement; the clays are used in the manufacture of Portland cement. The sands also yield phosphates, and fuller's earth is not uncommon; there are beds of white sand used in glass-making, and some very hard beds have been found suitable for millstones. In the West of England the Lower Greensand yields some rich iron ore; similar deposits have been worked also in Bedfordshire and elsewhere.

This formation is, from its great permeability, its

continuity and extent of outcrop, reliable as a source of water-supply; indeed, it was at one time proposed to supply London by deep borings into the Lower Greensand. If the series of permeable strata of which it consists had been proved to exist (as believed) in absolute continuity beneath the metropolis, no more constant source of good water need have been desired. But it is not so; there is a ridge of older rocks, buried beneath the Secondaries, standing up sufficiently high to cut out the Lower Greensand, which occurs only in thin patches, if at all, beneath the area where the supply was required. The existence of this old ridge has been established only by recent borings, and could not have been suspected merely from a survey of the outcrop of the Cretaceous rocks, which is unbroken all around the district which they occupy. The beds, no doubt, thin out gradually against this elevation, and where they occur in any thickness, invariably yield, when pierced, a good supply of water; resting, as they do, on the S. side of the ridge upon the Weald clay, and on the N. upon a clay of either Oxford or Kimeridge age.

The *Wealden* series consists of alternating clays and sands with intercalated bands of limestone, ironstone, sandstone and conglomerate. The Weald clay is used for making bricks; near its base are thin bands of limestone known as 'Sussex marble,' and it encloses some layers of nodular clay-iron ore which have been dug for commercial purposes. There are also beds of calcareous grit quarried for building and road-mending; one bed of calcareous sandstone, being very fissile, is employed for roofing and paving. The beds include some coarse and friable sandstones, not of a durable nature, some hard calcareous and ferruginous sandstone used for road metal, beds of shelly limestone and conglomerate used for roads and for rough building purposes. These beds also enclose layers of ironstone which have been worked for smelting.

The Lower Cretaceous soils vary rapidly according to the change from clays to limestones or sands, some of them yielding a soil peculiarly adapted to hop-growing; the clays are wet and more adapted to grazing, the lighter soils to agriculture.

The water-yielding properties of the Wealden beds are uncertain, on account of their frequent variation and lithological peculiarities. Water can be obtained almost everywhere within the area occupied, but it is at a considerable depth in many places, owing to the thickness of the Weald clay—this deposit has, however, some subordinate water-bearing beds. The Weald is especially an area in which, prior to commencing borings for water, the local geological structure should be subjected to thorough investigation.

The following remarks occur in connection with water from deep wells in Cretaceous rocks, in the 'Report of the Rivers Pollution Commission, 1868,' pp. 99, 102.

'Both with regard to the quality and quantity of the deep well waters which they yield, these formations (the Greensand and the Wealden) are considerably inferior to the New Red Sandstone, Oolite and Chalk. The Chalk constitutes magnificent underground reservoirs in which vast volumes of water are not only rendered and kept pure, but stored and preserved at a uniform temperature of about 10° C. (50° F.) so as to be cool and refreshing in summer, and far removed from the freezing point in winter. It would probably be impossible to devise, even regardless of expense, any artificial arrangement for the storage of water, that could secure more favourable conditions than those naturally and gratuitously afforded by the Chalk, and there is reason to believe the more this stratum is drawn upon for its abundant and excellent water, the better will its qualities as a storage medium become. Every 1,000,000 gallons of water abstracted from the chalk, carries with it in solution on an average $1\frac{1}{4}$ tons of the chalk through which it has percolated, and thus makes room for an additional volume of about 110 gallons of water. The porosity or sponginess of the Chalk must therefore go on augmenting, and the yield from wells judiciously sunk ought, within certain limits, to increase with their age.

'The only drawback to these waters is their hardness, but this disadvantage is greatly reduced by the circumstance that it is chiefly of the "temporary" kind, and can be therefore easily and cheaply removed.'

Upper Oolites.

The *Purbeck Beds* form a series of clays and limestones, the most important being the Purbeck Marble, which occurs in thin beds as a compact grey limestone, made up almost entirely of univalve shells, and formerly much used in the internal decorative work of churches. A softer limestone, which is capable of resisting the action of fire, is found in these beds, and is known as the 'burr-stone;' some quarries yield thin slabs suitable for roofing purposes.

The *Portland Beds* include the well-known Portland stone, of which St. Paul's Cathedral is built, and which is largely used for stone stairs and other internal domestic work. It is a white somewhat oolitic limestone, enclosing shells, and it occurs in beds ranging even up to 15 feet in thickness, but varying in hardness according to locality; some of the softest beds are used for holystone. Beds of phosphatic nodules occur, and are worked in some localities where the Portland beds are represented. These strata yield a poor and brashy soil.

The *Kimeridge clay* is a bluish-coloured shaly clay with occasional beds of bituminous shale. The clay is more or less calcareous, and is sometimes used in brick-making; it contains layers of septaria, and argillaceous iron ore has been found near its base. The bituminous shales have been burnt as fuel and distilled for gas and mineral oil.

The Purbeck and Portland beds yield a poor and brashy soil; the Kimeridge clay forms a cold, stiff and not very productive soil, more useful as grass- than as arable-land. The Upper Oolites, above the Kimeridge clay, generally yield water of good quality, the supply, of course, varying according to the local conditions; these may be greatly affected, in such strata, by faults and fissures, especially in the thick-bedded Portland stone.

Middle Oolites.

The *Coral Rag* or *Coralline Oolite,* occurs as a rubbly oolite and clay, between the Upper and Lower Calcareous Grit, where these members of the same series are present. It is an earthy calcareous freestone, sometimes used in building, but is not of a durable character; the Calcareous Grits include beds of sand and sandy limestone. The Coral Rag contains some beds of oolitic iron ore, which have been worked in Wiltshire.

The *Oxford Clay* is calcareous, dark-blue in colour, and is, in places, largely worked for brickmaking; it contains much iron pyrites and many bands of argillaceous limestone nodules; at its base occurs, over a large extent of country, an irregular calcareous sandstone, which in some parts is used for building. It forms a retentive soil, productive in some localities, but there is risk in its cultivation; it answers well as pasture-land. The Coral Rag and Calcareous Grits make soils which are light, brashy and far from being productive.

These beds do not yield much water, except locally, where the Grits are of a sandy nature, that from the Oxford Clay (in which water is sometimes found) being generally too impure for domestic purposes.

Lower Oolites.

The *Cornbrash* is the only constant member of this division of the Oolites, a series which assumes a totally different character, even in adjacent localities, or rather the beds rapidly thin out, and are replaced by others. This is a coarse, earthy, blue limestone, generally weathered, where seen in quarries, to a pale colour and rubbly condition. It is seldom good enough for building, but is used as road material, and occasionally burnt for lime.

The *Great Oolite* limestones are generally overlaid by a considerable thickness of clay, referable to the same period, and, in the South-West of England, also by the Forest marble, a fissile oolitic limestone, associated with thin beds, used as flagstones, for farm buildings, and roof coverings. The thicker slabs are useful for building rough ashlar work in even courses, and the material is of good quality for road-mending.

The Great or Bath Oolite series consists of pale yellow freestones, finely oolitic and free from fossils, with bands of shelly and argillaceous limestones. The Bath-stone is blue in colour far below the surface of the ground, and is soft when first quarried, but it hardens and changes to a yellow colour on exposure to atmospheric influences.

The *Stonesfield Slate* consists mainly of two beds of calcareous sandstone, associated with shelly, oolitic, and sandy limestones. The sandstone beds are so fissile after exposure to frost that they split up into 'slates,' much used for roofing; the beds also yield, in some places, a freestone that is quarried for building.

The *Collyweston Slate* is a similar calcareous sandstone, also largely quarried for roofing purposes. Intervening, geologically between the two, in Lincolnshire, occurs a thick series of marly limestones, known as the *Lincolnshire Oolites,* which afford excellent building stone, and are also burnt for lime.

The *Inferior Oolite* is darker in colour than the Great Oolites, and comprises sandy limestones, with some beds more compact in texture. The freestones are largely quarried in some localities for building, but are not generally valuable; they are soft when first quarried, and harden on exposure. In Yorkshire is a series of beds representing the Lower Oolitic period, and comprising limestones and sandstones, used for building, with beds of oolitic ironstone, which have been worked for iron ore.

The *Northampton Sand* is worthy of note as including

a valuable bed of sandy ironstone, often several feet thick, and which yields 50 per cent. and upwards of pig iron. The bed as an iron ore is strictly local, sometimes rapidly diminishing to a few inches only in thickness, or disappearing altogether. The series includes also subordinate beds of sandstone, quarried for building, and clay beds sometimes, although rarely, pure enough for the manufacture of terra-cotta.

The Cornbrash is 'said to derive its name from the facility with which it disintegrates and breaks up—brashy—for the purposes of corn-land.'—*Page.* 'As the name implies, the soil . . . is well suited to the growth of corn. According to Professor Buckman, it contains more phosphate of lime than the subordinate oolitic formations.'—*Woodward.* The loose and brashy soils of the Forest Marble and the Great Oolite limestone are poor and unproductive; those on the clay are better, but not by any means rich or fertile. The Lincolnshire Oolites make a light soil, sometimes red in colour, easily worked, but naturally not very productive. The brashy soils derived from the Inferior Oolite is poor on the higher ground, but in the valleys it is fairly fertile, as are those formed from the disintegration of the Yorkshire Oolites and the Northampton Sand.

The Lower Oolites are, from their composition and structure, well adapted to the distribution of underground waters; they may generally be relied on as a source of supply, and in certain favourable positions they throw out some of the finest overflowing springs to be found in this country.

'Unpolluted spring water from the Oolites is unsurpassed in its comparative freedom from all kinds of organic impurity. It is clear, colourless, palatable, and wholesome, and fit for all household purposes except washing, for which it is too hard. It may, however, always be softened by Clark's inexpensive process, and it then unites all the qualities which are most desirable in water supplied for domestic use. The Oolitic rocks are very porous, absorbing and holding enormous volumes of water, which are again delivered as springs, usually of great size. As water-bearing strata, or as a subterranean reservoir for the purification and storage of water, the Oolitic rocks are equal, if not superior, to the Chalk itself. But this vast store of magnificent water is rarely supplied to communities until it has been hopelessly fouled in river channels by polluting matters of the most disgusting description.

'The Oolitic rocks consist almost entirely of carbonate of lime; and this substance being soluble in water containing carbonic acid, springs issuing from the Oolite always contain a large proportion of solid impurity, of which the most abundant constituent is carbonate of lime, the remainder consisting almost entirely of mineral saline matter, also not injurious to health.—'Report of the Rivers Pollution Commission,' p. 120.

LIAS.

The *Upper Lias* is a blue shaly clay enclosing layers of septaria and nodules of bluish limestone, and in its lower beds jet is of frequent occurrence. The clay is much used for brickmaking, some of the beds being so bituminous that they burn with little or no fuel; and the more shaly beds, which contain iron pyrites in large quantities, for the manufacture of alum.

The *Middle Lias,* or *Marlstone,* consists of micaceous finely laminated clays and sands, with marls and an agillaceous and ferruginous rock-bed, which frequently forms a valuable iron ore. Where the rock-bed occurs as a limestone it is used for roads and building, and is occasionally pure enough to be burnt for lime.

The *Lower Lias* consists of clay, with many beds of blue and grey argillaceous limestones in its lower portion, valuable as building and paving stones, and from which good hydraulic lime is made. Layers of septaria also occur, which are used in the manufacture of hydraulic cement; the clays are dug for brick-making.

The *Rhœtic Beds*, at the base of the Lias, include several white limestones, used for lime and building purposes, 'capped by a hard, smooth-grained stone, called the "sun bed," which from its closeness of texture and general purity has been recommended for the purposes of lithography. At or near the base of the White Lias is found the Cotham or Landscape Marble.'—*Woodward.*

The Upper Lias soils are good as grass-land; the Middle Lias beds form a rich soil 'favourable to the growth of apple trees;' that derived from the rock bed is frequently red in colour and equally productive. The soil of the Lower Lias, although brashy in parts, is fairly fertile, and forms excellent pasture and dairy land. Some grey marls in this Rhaetic series are useful for marling land; the soils it makes are of good quality.

The Marlstone rock-bed yields water, in some localities only; as do the limestones of the Lower Lias, but in quantity not to be relied on, or in quality to be recommended.

Trias.

The *Upper Keüper* series consists of variegated

marls with occasional beds of sandstone, common gypsum and alabaster. The marls are dug for brick-making, the sandstones quarried for building, the common gypsum is burnt into plaster of Paris, and the purer kind, or alabaster, is used for ornamental purposes. Fuller's earth occurs in the marl, and patches of rock-salt in the marl and sandstone; very thick beds of the latter mineral are found in Cheshire and adjoining counties, and are extensively worked by mining.

The *Lower Keüper* sandstones, called also water-stones, form a thick series of micaceous red and whitish sandstone, with a base, in places, of hard dolomitic conglomerate. The uppermost beds are generally finely laminated, those in the centre are good freestones employed in building. The lower beds afford road material, and in some places are sufficiently calcareous to be burnt for lime, in others they yield copper ore. In the South-West the Triassic Rocks are magnesian in character, are quarried for lime and for building purposes, and contain workable beds of iron ore, also clays which are dug for brickmaking.

The *Bunter*, or *New Red Sandstone*, consists essentially, as its name implies, of red or reddish sandstones, sometimes variegated, and of different degrees of hardness and suitability for practical purposes. At the base are pebble beds, loose or cemented into conglomerate, which in places yield ores of lead and copper.

The Upper Keüper marls form a rich soil well suited for orchards and pasture-land. The lower beds of the Trias make generally a poor sandy soil.

The Upper Keüper seldom produces large springs,

and the water is frequently charged with salt; but the Lower Sandstones form an excellent water-bearing formation, being permeable and fairly persistent. The beds of the Bunter also yield water; indeed the Triassic rocks, as a whole, are second to none in value as sources of supply to the deep-seated springs.

'The New Red Sandstone rock constitutes one of the most effective filtering media known, and being at the same time a powerful destroyer of organic matter, the evidence of previous pollution, in water drawn from deep wells in this rock, may be safely ignored, unless the previous animal contamination has been very great indeed.

'The unpolluted waters drawn from deep wells in the New Red Sandstone are almost invariably clear, sparkling, and palatable, and are among the best and most wholesome waters for domestic supply in Great Britain. They contain as a rule, but a moderate amount of saline impurity, and either none, or but the merest traces of organic impurity. The hardness is usually moderate, and only when the water is derived from originally impure sources does it become excessive. There is every reason to believe that a vast quantity of hitherto unutilised water of most excellent quality is to be had at moderate expense from this very extensive geological formation.'—'Report of Rivers Pollution Commission,' p. 94.

CHAPTER III.

MATERIALS, MINERALS, AND METALS—*continued.*

Economic products of the Palæozoic Rocks.

PERMIAN.

THE *Permian* or *Magnesian Limestone* series consists of red and white sandstones, magnesian limestone, and variegated gypseous marls. The chief of these is a yellowish limestone, composed of nearly equal parts of carbonate of lime and carbonate of magnesia. Several varieties of this limestone occur, some being laminated, others oolitic; it is generally a good building stone, but it is important to remember that the beds vary greatly in durability. Some hard flinty beds are used for road-mending; the marls have been dug for making bricks, and in places they contain beds of gypsum. The red sandstone beds also are much quarried for building, and the sharper varieties for whet- and grind-stones; some of the associated basalts are largely used as road material.

These beds produce a light soil of reddish colour, and, on the marl especially, of good quality.

The permeable sandstone beds are water-bearing, and the limestones also under certain conditions—which, however, are local, and demand examination.

Carboniferous.

The *Upper Carboniferous* series consists of rocks of greatly varying character; sandstones, including the York stone, grits, conglomerates, shales, ironstones, and coal. The sandstones generally are quarried for building, the flagstones for paving and, when thin, for tiling purposes; some sandstones and grits make exceptionably good mill- and grind-stones. The clays are employed, some for making bricks, pottery, and earthenware, others for fire-bricks and encaustic tiles. The shales yield iron pyrites, which is used in chemical manufactures, and the beds of clay-iron-ore are well-known for their extent and productiveness. This series includes also the vast and extensively worked coal measures of Great Britain.

The *Lower Carboniferous* beds are of more calcareons nature, consisting of argillaceous limestones, crystalline limestones, dolomitic in some localities, oolitic in others, calcareous sandstones, clays, and conglomerates. There are some beds of micaceous sandstone quarried for paving and building. 'The carboniferous limestome is much quarried for lime. It is largely used for rough buildings and for road-mending, for which purpose it is conveyed to great distances, but from its hardness it is not serviceable as a freestone. Some of the beds are polished and used as marble for ornamental purposes.'—*Woodward.* Basalts occur, and a fragmentary volcanic ash in Somersetshire is worked for building material; bordering the Pennine and the Cheviot Hills are large masses of basalt and porphyry, and granite occurs in

Cornwall and Devon. In the Carboniferous limestone are many dykes and veins containing ores of zinc and lead, and many valuable beds of iron ore belong to this series.

The Upper Carboniferous sandstone soils are barren, those of the clays and shales are more productive. The Carboniferous limestone makes a poor ferruginous soil, in some parts so thin as to be fit only for sheep pasture.

The water from the Coal measures is not, as a rule, of good quality, that from the Millstone Grit is better, and can generally be obtained by boring. The Lower Carboniferous cannot be relied on as water-bearing strata, especially the Mountain limestone.

Devonian.

This system of rocks comprises many hard sandstones, red or grey in colour, with conglomerates, slaty beds, shales, and occasional limestones. The slates vary greatly in quality, those quarried in Cornwall being the best found in this series; in some places they pass into useful hone or whet stone. Greenstone is worked, and basalt occurs in Cornwall and Devon; also granite, with granitic dykes of a more felspathic rock, called Elvans, useful as road metal, aud which forms a durable building material. The decomposed granite forms kaolin, a valuable china clay. The Devonian rocks are traversed by veins which yield ores of silver, copper, lead, zinc, tin, manganese and iron.

A poor soil is formed by the surface decomposition of the Devonian rocks, which, owing to their thickness and permeability, are, in some localities, good water-bearing strata.

Old Red Sandstone.

As suggested by its name, this series is mainly arenaceous; it comprises micaceous sandstones, red, grey and mottled, conglomerates, marly shales and slates, with local beds of nodular limestone known as 'cornstones.' The sandstones are used in building and road-mending, the cornstones also for the latter purpose, and the conglomerates for rough millstones; some of the beds are firestones.

The Old Red Sandstone yields a loamy fertile soil, suited to the growth of hops and apple-trees; the cornstones also make very rich land. The formation yields large quantities of good water.

Silurian.

The *Upper Silurian* rocks consist mainly of hard siliceous sandstones, sometimes micaceous, grits and conglomerates, some of the bands being calcareous. It includes also nodular limestones burnt for lime, thin shelly limestones, calcareous flagstones, marls, shale and slates.

The Silurian rocks cover very large areas, and produce soils, frequently red in colour, of varying character; the water-bearing characteristics of this great system of rocks are equally various, according to locality and physical conditions.

Cambrian.

This group of rocks—which includes those formerly called *Lower Silurian*—occupies an extensive area and furnishes much valuable building material. Its chief economic product is roofing slate, varying much in quality and appearance, and rendering the ground where that of the best kind occurs of enormous value. The world-renowed quarries of Llanberis and Penrhyn are notable examples, and the green slates of Bangor, on the same geological horizon, are much sought for ornamental purposes, as are those of the same colour, which occur as part of a higher group in the lake district of Westmoreland. The blue and dark slates of Dolgelly are soft, and those of Skiddaw, although occasionally used, are unsuitable for roofing purposes, yielding readily as they do to the action of the atmosphere.

The good building stones abundantly yielded by the Cambrian rocks are sandstones of every degree of fineness and durability, some hard and compact, others soft and fissile—limestones, varying from impure earthy beds to fine marble—calcareous slates and flagstones, siliceous sandstones, freestones, quartzites and conglomerates. There are also hard and highly metamorphosed grits, sandstones and schists, attaining a great thickness, and including masses of intrusive greenstone. Dykes of greenstone occur amidst the slates of Llanberis and Penrhyn, and near Caernarvon is an intrusive mass of tough felsite, which is valuable as material for paving. The granite and syenite of Charnwood or Charnley Forest are largely quarried for road metal, and the Whittle-hill oilstones are obtained from the metamorphic rocks of the same locality, where the Cambrians occur as an inlier. Welsh oilstones come from the Cambrian beds near Llyn Idwal, and Cutlers Greenstone is worked on Snowdon. The beds yield, at Dinas-Mowddwy, a sand which is used for lining copper furnaces, and 'some valuable deposits of phosphate of lime—phosphorite—have been discovered on the top of the Bala limestone in North Wales.'—*Woodward.* The Cambrian rocks in some parts yield ores of copper, lead, zinc, iron and cobalt.

Laurentian.

The Laurentian rocks occupy but a very small portion of the surface of these islands, and are generally included in the same colour on geological maps as those of the Cambrian system, which they underlie. They consist of gneiss and gneissose rocks in the island of Lewis, and of sandstones and conglomerate in Sutherlandshire.

CHAPTER IV.

SPRINGS AND WATER-SUPPLY.

Nature of Springs—*Surface Springs—Deep-seated Springs—* Water-level.

Nature of Springs.—All sources of water-supply depend entirely upon physical conditions—the rainfall upon situation and elevation, the rivers upon physical features, and the springs upon geological characters. As physical conditions are more or less different in every district, the sources and available quantity of water within them also vary, being wholly governed by those conditions. The physical features have, in addition, a direct and important bearing upon the methods of distributing the supply, in the quantity and manner best suited to the requirements of each locality.

'It may be taken as a general rule that ridges and escarpments consist, wholly or in part, of water-bearing beds, such as limestone, chalk, or sandstone, and that the softer clays, being more readily denuded, occupy the lower grounds.

'Strata generally dip towards and pass under the higher grounds; it is rarely they occur otherwise, and they are not often found quite horizontal. This is owing to the fact that the lie of the strata has, in a great measure, given to the ground its present form; in other words, anticlinals are more easily denuded than synclinals, the latter remaining whilst the former have yielded to denudation.

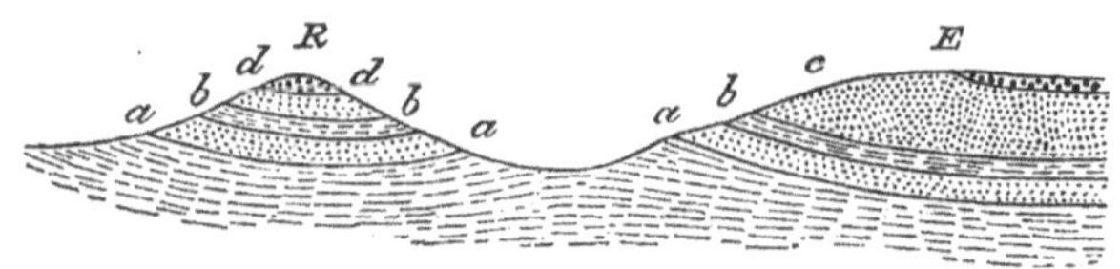

Fig. 7.—Section of a ridge and escarpment of pervious and impervious strata, to illustrate the nature of springs and their water-level.

'*Case* 1.—Any set of pervious strata which occurs at the surface, as at *a a*, in figure 7, will (as a rule) form a trough, and be found below the ridge, R, at a greater depth than is due to the difference in height of *a* and R. Water collected by these pervious beds upon their outcrop, *a a*, will rise in a boring made at R, through the impervious beds, *b b*, to a height corresponding (or nearly so) with that of their outcrop. 'Surface gravels are no exception to this general rule, and often occur in long hollows on elevated ridges, throwing out intermittent springs at the lowest points along their margin, *d d*.

'*Case* 2.—The same pervious beds, *a a*, when dipping into the face of an escarpment, E, probably decrease in dip at a short distance beneath it, or even become horizontal, and they would be reached by a boring made on the top of the escarpment, E, at a depth not much below that of their outcrop, the water from them rising in the bore to an extent coinciding with the difference, whatever it may be.

'*Case* 3.—Where pervious beds, *c*, in considerable thickness, overlie impervious beds, *b*, and form an escarpment, E, or a ridge, R, the water-level within them rises beneath the higher grounds until the forces of hydrostatic pressure and frictional resistance are in equilibrium. Water will, therefore, be frequently found in borings commencing on a ridge, R, or an escarpment, E, at a height very considerably above that of the lower points of their nearest outcrop, *b b*, where the surplus water flows forth as perennial springs.'*

* Extracted from an essay on 'National Water-Supply,' by the author. *Journ. Soc. Arts*, July, 1879.

As the rain falls upon the ground it is at once drawn by the force of gravitation to lower levels. If the surface be impervious, the water runs off by the ditches, rivulets, and rivers to the sea; but if it be wholly or partly pervious, a portion only thus flows to the ocean, the remainder passing into the water-bearing formations. The water which is thus absorbed is again thrown out, at a lower level, in the form of springs on the hill-sides, or, the conditions being favourable, it is retained within the strata at various depths, thus forming huge subterranean reservoirs. These stores of water, being supplied from large collecting areas, and replenished by every shower of rain, are, in the great majority of instances, almost if not quite inexhaustible, and not only from the quantity, but also from the quality of their waters, merit a greater share than they have hitherto received of practical attention.

Surface Springs.—Although all springs owe their existence to the same hydrostatic principle, they are found in many different forms, and under greatly varying conditions. The simplest kind of spring is that in which the water occurs over a definite area at or near the level of the ground, and is, therefore, called a surface-spring; it consists of a body of water held up in a superficial stratum of gravel or sand, by a bed of clay or other impervious rock beneath. The water overflows at the lower points along the upper edge of the impervious bed, thus forming springs which flow more or less rapidly according to the seasons, and which, if the collecting ground is small, may be intermittent.

Where such conditions exist the bed of gravel or

sand is of course saturated up to a certain level, and any wells dug down thereto meet with land or surface-springs, which yield a more or less plentiful supply.

But it should be borne in mind that these springs are fed by rain and surface water which has passed down through a filtering permeable rock, the thickness of which is probably insufficient for the elimination of suspended matter and the oxidation of organic impurities; also that the permeable rock itself is liable to be charged with the elements of contamination, for its spring-producing permeability is a great source of this danger. Imperfect drains and leaking or overflowing cesspools within it, farm-yards, and refuse-heaps upon the surface, must each contribute their share of pollution, in the form of decaying organic matter, if not of that which may be of even more serious nature.*

The supply from surface springs will vary directly as the seasons; in periods of drought the streams run dry, the springs fall off, and shallow wells, which derive their water either by soakage from a stream or from a so-called land spring, become exhausted in consequence. Neither streams nor land-springs should ever be depended on for a supply of water for domestic purposes, not only on account of their intermittent nature, but also because of their liability to contamination.

There are also surface springs, in which the primary geological conditions are reversed, the water-bearing

* See the 'Report of the Rivers Pollution Commission,' 1868; under headings 'Shallow Well Waters,' p. 168; and 'On the Propagation of Epidemics by Potable Waters,' p. 140.

bed of pervious sand or gravel being beneath a thin covering of impervious clay. In these cases, the springs of one locality are fed by water from another at a slightly higher level, but at no great distance, from which it travels, at a small depth only, beneath the surface. The water-bearing bed is thoroughly saturated, in this case also, to a level varying with the seasons, and it yields water wherever penetrated below that level, but which is also liable to be locally affected by surface contamination; but it may be that the higher part only of such a stratum is polluted, for at a lower part, if at any distance from the upper, some of the injurious particles may be found to have undergone a chemical change, and others will be removed by filtration. Springs of this kind are to be preferred to those first mentioned; they may be intermittent, but are more constant in their flow, and more to be relied on in respect of purity; they are intermediate in character between the simpler surface springs and those now to be described, the supply to which passes at a considerable depth, and from a greater distance.

Deep-seated Springs.—These springs, although properly called deep-seated, are not necessarily at a great depth beneath all the area under which they occur, for their waters will sometimes even flow out again at the surface, if the beds along which they pass be anywhere exposed at a level lower than that of the outcrop upon which they were collected. But they will have been deep-seated for a greater part of the distance traversed by their waters, owing to the inequality of the ground, the dip of the water-bearing beds, or to both causes combined. And even when they approach or reach the surface, after a long underground passage, they differ from surface-springs in being much more constant in the yield and equable in the temperature of their waters, which have been also freed from organic impurities by oxidation and filtration.

'Surface-polluted water, when it penetrates only to shallow wells, still retains a considerable proportion of its polluting organic matter in an unoxidised condition. But when it descends through 100 feet or upwards of porous soil or rock, the exhaustive filtration to which it has been subjected, in passing downwards through so great a thickness of material, and the rapid oxidation of the dissolved organic matters in a porous and aërated medium, afford a considerable guarantee that all noxious constituents have been removed, even from such portions of the water as have passed perpendicularly downwards. Still more so must this obviously be the case with the even much larger portion which reaches a well in a more or less horizontal direction, through far greater thickness of porous medium.'*

The supply of water to all springs is derived from the rain which falls upon, or flows off impervious strata to, the outcrop of the pervious strata in which they occur. The outcrop thus forms their collecting-ground, and the yield of the springs will be great or small according to the area occupied by the outcrop of the permeable beds, the rainfall thereon, its elevation, and the by no means constant degree of permeability of the rocks themselves. Other conditions, such as the existence of faults and fissures, and the thinning out of strata, affect the underground passage of the water, beneficially or otherwise, and need only to be mentioned as worthy of careful consideration in the

* 'Report of the Rivers Pollution Commission,' 1868, p. 89.

estimation of water-supply to any locality. The quality of the water from each spring will be governed by the chemical composition of the rocks it traverses, and the readiness with which these yield to the water, by chemical changes, their organic and inorganic constituents.

As many of the permeable rocks, which form elevated tracts of land receiving their due share of the rainfall, must in other areas, and at lower levels, form the floor of the sea, the fresh-water of their springs is constantly flowing directly into the salt-waters of the ocean. Such springs are known from which fresh-water is constantly drawn, and, on the other hand, sea-water traverses the permeable rocks and affects their springs, sometimes for many miles inland. Where the requisite conditions are found, of pervious beds dipping under the sea and covered by those which are impervious, fresh-water can always be obtained by boring to the deep-seated springs. Springs thus situated form a valuable source of supply to islands and places along the coast where the water from the surface-springs would perhaps be quite useless for drinking purposes, in consequence of its holding much salt in solution.

For much valuable and detailed information on this subject, the reader is referred to :—

The Sixth Report of the Rivers Pollution Commission (1868), 1874.

Water and Water-Supply. Ansted. (Allen.)

Water Analysis. Professor Frankland. (Van Voorst.)

Experimental Researches. Professor Frankland. (Van Voorst.)

Water-level.—In surface-springs, the water may be under various and even under varying conditions, and the water-level—*i.e.*, the plane of the upper surface of the water—may be constant or variable according to the dip of the beds, the position of the spring, the time of year, and the variability of the seasons.

The level of the water of a deep spring in pervious strata, which is kept down in them by overlying strata that are impervious, as in *a*, figure 7, will be constant at, or slightly below, the height of the nearest point of outcrop of the water-bearing beds. The water is under hydrostatic pressure varying with the depth of the beds below the level of the outcrop, and its level is but slightly modified by the frictional resistance offered to the passage of the water through the more or less compact material of which the beds are composed. Consequently the water will rise in a boring made down to such a spring to the normal water-level; this can generally be ascertained, for any spot, by calculation from data based upon the geological conditions considered in connection with the physical features.

In deep springs, where the water is held up in a great thickness of permeable strata by an impermeable stratum beneath, as in *c*, figure 7, and the pervious beds are not saturated throughout their entire thickness, the water-level is influenced by a different set of conditions in the following manner:—The water-level of the springs thus formed will be, at the margin of the pervious bed, coincident with that of their lower surface boundary, but not merely up to this level will they be saturated. In the area occupied by their outcrop, the water-level will be found to rise within them from the marginal line referred to in a ratio proportionate to the permeability of the beds and the amount of the local rainfall, and to fluctuate according to the seasons. Rising thus as it does from the lower boundary-line of the permeable beds, it will fall in a transverse direction from the higher ground towards each lateral valley by which they are intersected. The reason for this rise in the water-level beneath the higher ground may be thus explained. The rain-water which has fallen upon the outcrop of permeable strata so situated will have percolated downwards, not only until the saturation was complete to the level of their boundary, but will have accumulated within them above that level under the higher ground until it has reached a height at which the forces of gravity, capillarity, and frictional resistance are in equilibrium. It must necessarily be that the water-level in springs thus situated will coincide in the valleys with that of the local streams by which the contiguous beds are drained, and rise from them in every direction. It is highest under the loftiest hills, because, with equal slopes, these would occupy the largest areas, and the water-level retaining its normal degree of inclination would attain beneath the centre its greatest elevation.

There may be more than one, even several, deep-seated springs within a practical depth at any particular spot, according to the number of alternating pervious and impervious formations, as, for instance, in *a*, and *c*, figure 7, page 132. Each spring may have a different water-level, due to the varying heights of the outcrop of the beds along which their waters are borne, and to the natural drain upon them by streams, which lowers it along the courses of the valleys, thus preventing the normal accumulation.

As an illustration of the phenomena of deep-seated springs and their water-level, the area included in Sheet 7 of the Ordnance and Geological Surveys has been selected. It is coloured as a geological map, without drift deposits, in the frontispiece, and a section across a part of the area is represented in Plate ii. The names of the formations are engraved on the map, and their water-bearing characters are indicated by different tints.

The outcrop of the Chalk forms the highest ground, which forms a large collecting area of permeable beds, from which a large proportion of the rainfall passes in under the Tertiary formations towards the south-east. The Lower Tertiaries are partly pervious, but as the overlying London Clay is impervious, the water is held down by them and it, considerably below the level of the Chalk outcrop. A boring made through the Tertiary beds, sufficiently far to meet with a fissure in the Chalk, would tap its deep springs, and the water would rise (as in Case 1 or 2, p. 132) to a height nearly coinciding with that of the outcrop. The water-level would, however, be subject to this modification :—the area is traversed by the rivers Thames and Colne, which flow over the permeable Chalk for a part of its length, and lower the water-level within it considerably, as shown in the section. Where this influence ceases the water-level again rises towards the N.W. (as in Case 3, p. 132), being highest under the hills, and falling towards all the lateral or tributary valleys.

The section is drawn to an exaggerated scale, but

Plate 2.

Section across Plate 1 to illustrate the level of the water from springs in the Chalk.

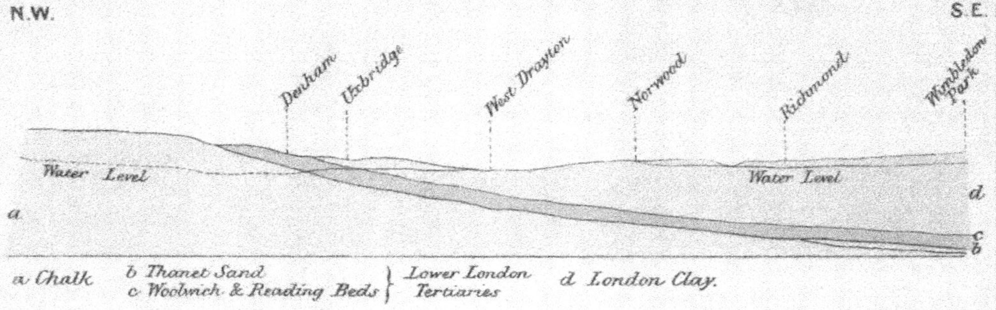

N.W.

S.E.

Denham

Uxbridge

West Drayton

Norwood

Richmond

Wimbledon Park

Water Level

Water Level

a

d

c

b

a Chalk

b Thanet Sand
c Woolwich & Reading Beds } Lower London Tertiaries

d London Clay.

the water-level is proved to be as described by several deep wells along its line; some of these are appended,* in the form usually employed for noting sections of deep wells and borings.

DENHAM.—The Tile House.

Dug 110 feet, the rest bored.

Water rose to 85 feet from surface.

		Feet
London Clay.	Yellow clay - - -	22
Reading Beds.	Gravel and sand, mixed - - -	15
	Gravel, sand and chalk, mixed - -	30
	To Chalk - -	67
	Chalk - - - - - -	128
	Total - -	195

UXBRIDGE.—'The Dolphin.'

Water rose to 3 feet from surface.

		Feet
	Soil, etc. - - - - -	4
	Gravel - - -	$13\frac{1}{2}$
London Clay.	Clay - - - -	20
Reading Beds.	Sands and clays - - -	44
	To Chalk - -	$81\frac{1}{2}$
	Chalk and flints - - - -	$39\frac{1}{2}$
	Total -	121

* 'Mems. Geol. Survey,' vol. iv. *The Geology of the London Basin*, Appendix (Whitaker).

WEST DRAYTON.—Victoria Oil Mills.

Shaft 12 feet, the rest bored.

Water overflowed.

		Feet
	Made ground - - - -	3
Valley Drift.	Brick-earth and gravel - - -	29
London Clay.	Blue clay, sand and pebbles at bottom -	88
Reading Beds.	Coloured clays and sands - - -	66
	To Chalk - -	186
	Chalk - - - - -	100
	Total - -	286

NORWOOD, near Hanwell.

Shaft through London Clay to sand, 280 feet.

Water rose to the top.

	Feet
To Chalk - - - - -	325
Chalk - - - - -	89
Total - -	414

RICHMOND.—'Star and Garter.'

Water rose to 39 feet from surface.

	Feet
To Chalk - - - - -	416
Chalk - - - -	76
Total - -	492

WIMBLEDON.

Shaft 200 feet, the rest bored.

		Feet
	Sand and gravel - - - -	10
London Clay.	Blue clay, sandy at bottom -	431
Reading Beds.	Mottled clays and sands - -	74
Thanet Beds.	Dark clays and greenish sands -	22
	To Chalk -	537
	Chalk - - - - -	30
	Total -	567

In the estimation of the available water-supply to any locality, from deep-seated springs, the formations beneath it are surveyed in the manner described in Part II.; a section is then drawn from the dip of the beds, ascertained from observation or from three or more points of outcrop, all probability of change in dip or permeability, of fracture and so on, forming part of the investigation. The probable depth to the water and the height to which it will rise can then be readily estimated, due allowance being made for the varying conditions, described in Cases 1, 2, and 3, page 132, in the resulting water-level. It is evident that the details required for this purpose are obtained, not at the spot where the boring is to be made, but upon the outcrop of the beds beneath, and frequently at a great distance from the point in question.

It must be borne in mind that the Drift deposits exert an appreciable influence upon the amount of water within the beds upon which they lie. A covering of Boulder Clay must exclude a good deal of water from an outcrop in some cases; in others it may even add to the amount received by those which are so situated as to catch what runs off from its surface. In some localities the Drift gravels yield large quantities of water; if they occur in any extent and beneath the Boulder Clay, it may be of good quality; but when they form the surface deposits, care must be taken to ascertain that their springs are not polluted by drains or other sources of surface contamination.

CHAPTER V.

SPRINGS AND WATER-SUPPLY—*continued.*

Artesian Wells—Absorption Wells.

Artesian Wells.—These wells take their name from Artois in France, where first used in western Europe, although they were employed in the East in very early times. They are borings rather than wells, and in their simplest form are only holes made down to deep-seated springs, from which the water rises in those springs to its normal level. The making of the bore-holes and the machinery employed are matters of pure engineering; but there are many problems in connection with such enterprises which demand prior geological investigation. These are, briefly, the thickness and hardness of the strata which must be passed through to reach a particular spring, the number of springs that may occur within a given depth, and the height to which the water from each will ascend.

Upon the hardness and thickness of the beds, and especially upon the frequency of variation in those particulars, must depend the cost of the undertaking, for it naturally makes a great difference whether a boring be made in a given thickness of rock of similar character throughout, or in the same thickness of rock in which many refractory bands occur. From this variation and other causes the bore may have to be repeatedly constricted as it descends; it has indeed often happened that a bore-hole, commenced with a too small diameter, has been carried a greater part of the way down to a spring, and then been of necessity abandoned. It is but seldom, although it is sometimes the case, that Artesian wells are commenced on absolute speculation, or without some idea of the source of the springs whence their supply of water is to be derived.

If water from a certain series of beds be desired, the depth to them, and their local nature, may be ascertained by accurately surveying their outcrop; of course when there are other borings or wells in the immediate vicinity, particulars of these will facilitate the inquiry—the outcrop may be near or distant, but there and there only, in the absence of other wells, is the requisite information to be gained. The main points to be ascertained, in the manner described in Part II., are the height of the outcrop in relation to the spot where the boring is to be made, the width occupied by the beds at the surface, their permeability, the angle of their dip, and its direction. From these data, if correctly observed, can be calculated the depth at which the beds will be met with in the boring, the height to which the water will ascend, and, within certain limits, its quantity and quality. An estimate can then be made of the cost, which must be based upon the hardness of the beds, their variation, and the depth of the spring below the scene of operations.

There may be, as we have seen, several beds or

series of beds, each producing springs of similar or of different character; these will be at various depths, and their waters may stand at different levels, but the method of calculation applies to all alike, and that spring which is most suitable can be selected. In passing through the upper beds yielding springs, precautions will of course be taken to exclude their waters from the bore-hole where that from a lower spring is, for any reason, desired. If the water from a lower spring be estimated to stand at a higher level than that from an upper one, and the latter be not excluded, the water—assuming the yield in each case to be equal—would not reach its own level, but would be absorbed by and would remain at that of the upper spring. By the same rule, if a boring be continued, after reaching a spring, until it pierces beds yielding water having a lower water-level than that of the spring first tapped, the water from the upper spring will stand at the level of that of the lower. Or, the lower beds may be permeable but yield no water, and the conditions affecting them such that if they did yield water it would stand at a certain level; then the water from above would be entirely lost in them until they were saturated, if such were possible, up to that level, throughout their entire extent.

'In forming plans for an artesian well, the first point is to determine exactly the size and depth necessary to obtain the requisite number of gallons of water per day, or rather the maximum number of gallons at any portion of the day in order that a reserve may be provided if that quantity exceeds the probable yield of the spring. Provision may be made for this by having a shaft of sufficient diameter and depth below the water-level to contain a quantity of water above and beyond what rises from the spring, equal to a difference between it and the demand during the period of largest draught. This reserve being exhausted, or nearly so during that period will be replenished in the intervening periods, when pumping is carried on at a rate less than that of the flow of the spring, or when it ceases altogether. Should a very large reserve be required, headings or chambers may be driven horizontally from the shaft, and these will increase to any extent the space available for accumulation. In certain cases such headings driven in the direction from which water may be found weeping into the well will increase the supply, but care must be taken that the additional water thus obtained is of a quality to render its use desirable. If not pure, it must be excluded by cylinders or other means, and the chambers will then be driven in a different direction, or in some stratum which is quite impermeable.

'It will next be necessary to decide whether an ordinary well dug down to the spring is to be preferred, or a shaft for a portion only of the distance supplemented by a boring. The decision will depend entirely upon the depth to the water-bearing beds and the level at which the water will stand, the circumstances which may influence the necessity for a reserve being at the same time taken into consideration. If a boring be decided on the shaft should be carried down, where practicable, several feet below the water-level, even if the pumps be not fixed below it, as the supply, when drawn from a body of water in a well, is less likely to be turbid than when taken direct from a boring having a much smaller diameter.

'In some cases it is desirable to put down a small trial-bore; for instance, when the indications of depth or water-level are exceptionally obscure, or where they are known to vary rapidly within a small area. The smaller hole is made at much less cost than a serviceable bore, and sometimes it may prevent a fruitless larger expenditure, but the necessity for such trials forms the exception. There is, however, one set of conditions in which their use is recommended—where springs are known to exist with a possibility of their waters being salt from having passed through beds of rock-salt. Such beds are frequently local and of small extent, affecting certain springs only, those at a contiguous spot holding no saline ingredient in solution.

'The kind of material best suited to the purpose of lining the shaft is governed entirely by the nature of the rocks through which the well passes. Some strata are coherent and will stand like a wall for any length of time ; some, although solid, are very liable to cave in ; others are of a crumbling texture and must be supported, even during the execution of the work—the solidity and stability of all, whether hard or soft rocks, may be affected by the presence of water within them. Each case must, therefore, be ruled by the details of the strata through which the well is to be made—in some, iron cylinders will be required for the whole depth of the shaft ; in others, for a part only will be sufficient. In rocks which are fairly dry and firm bricks may be used for lining the well with or without a coating of cement inside, and in others (although these form the exception) no lining whatever is actually necessary.

'The expense of sinking or boring for water is not always proportinate to the hardness of the rocks to be penetrated. Frequently those which are the most compact in a hand-specimen, are the most readily pierced, owing to their being jointed in several directions ; whilst, on the other hand, some sands which may be ground into dust between the fingers, are so tough and coherent in the mass that they have to be picked to pieces with chisel and hammer, or even blasted with gun-powder. Sometimes sands full of water, termed "quicksands," are met with ; these demand special precautions and perhaps considerable outlay, if it be necessary to pass through them and to exclude their water.

'There are cases in which tube wells may be advantageously employed at a cost much below that of artesian or even of ordinary wells. Although chiefly adapted and generally used for obtaining a supply from surface springs, they have been successful in tapping those at a considerable depth. In a loose material, such as sand, the pumping from a tube-well forms and gradually enlarges a cavity around the base of the pipe which acts as a small underground reservoir. Assuming the cost of the tube-well pipes to be the same as of those used for lining a bore-hole, any saving arises from the difference between the cost of boring and that of driving, and under certain conditions this is material; but a tube-well can be successful only where the strata are throughout such as will admit of the pipes being driven through them.

'A plan has been highly recommended which may be described as intermediate in character, combining the peculiarities of the tube and artesian well, applicable wherever water is to be obtained, and independent of the hardness of the rocks. It is to attach a pump to the pipes lining either a tube-well or a bore-hole, at a certain distance above the water-level, and without a shaft for the accumulation of water. By constructing an air-tight chamber between the pump and the surface of the water, it is affirmed that the supply from a low spring may be considerably increased, whilst the water is thereby freed, at the same time, from much of the matter it may hold in suspension.

'All wells and borings, of whatever kind, should be carried down to the springs which yield the most copious supply of good water; when, however, two or more springs exist at a workable depth, and there is no appreciable difference between them in regard to yield or quality, that one should be chosen from which the water will stand at the highest level. This is a very important point, upon which depend the depth, size and cost of the shaft in ordinary and artesian wells, and, in a great measure, the cost of the pumps and pumping the water to or above the surface. The distance down to the spring, not to the water-level, must also determine, in connection with the nature of the rocks, the size of a bore-hole at its commencement. A bore may continue to a great depth unaltered in size where no change occurs in the character of the strata, but when soft rocks alternate with those which are hard it has to be frequently constricted; therefore, unless a bore-hole be begun of sufficiently large size, it may, of necessity, be decreased to a size at which it cannot be continued down to the point it was intended to reach. The bore should be at starting of a sufficient diameter for this reason also; if it has to be contracted many times owing to hard rocks or other impediments, although it may eventually reach the spring, it will otherwise be too small to yield water at the rate which may be necessary even if the spring itself is capable of affording the requisite supply.

'There are several ready methods of testing the yield of a spring, one being the delivery into tanks of so many gallons of water in a given time from pumps capable of lowering the water-level. Another is to note the rate at which the water fills the shaft after having been pumped down, the calculation resting on the size of the shaft and the number of seconds or minutes it may take for the water to regain its normal level in the well.'*

Absorption Wells.—The phenomenon mentioned on page 147, of one spring absorbing the water from another suggests a valuable, although not well-known, practical application. It has been stated that strata which now throw out springs would, if occurring at a different level, or if inclined at a suitable angle, become the means of draining water away from the surface. This would occur as a matter of course, with reversed conditions, for the water would merely be passing through the same beds in another direction; the springs of one locality are in fact but the natural drainage of another.

The proposition might not at first be readily accepted, but a moment's thought convinces of its truth, that a well which is capable of yielding a given quantity of

* Extracted from an article on 'Pure Water and its Sources,' by the Author, in *The Brewers' Journal*, 1879, in which are given simple directions for the qualitative analysis, and for the determination of the hardness, of natural waters.

water is equally capable of getting rid of the same quantity by absorption. Water is not elastic like a gas, and when it bursts up from a penetrated rock does so not from the force of expansion, but simply from that of hydrostatic pressure. And it rises to a certain height only, that of its normal water-level; when this is attained in the well, and there is no outlet below that level, no more water rises. Of course as it is pumped out more takes its place, if the pumping does not exceed the yield of the spring, but it never stands higher than a definite point. Therefore, if water be put into the well, for the moment tending to increase the height above that point, the pressure is reversed, and the water so put in sinks down to the water level, at once if its quantity in this case also do not exceed the yield of the spring.

The rapidity with which water flows from a spring into a well depends directly upon the permeability of the strata through which it has passed, and upon the same conditions depends also the rapidity with which it can flow away from it; consequently the spring is capable of yielding and of absorbing the same quantity of water. This power of absorption has been unwittingly made use of in thousands of instances, and in some with most disastrous results; the pollution of surface springs by drains and cesspools needs only to be mentioned. In this way a very valuable source of water supply for domestic purposes has been utterly destroyed in the gravels upon which almost all the old towns and villages are built, and traces of sewage are recognisable even in many deep-seated springs. There are, however, cases in which absorption wells may be employed with safety and advantage, for instance in getting rid of a troublesome excess of water from surface gravels, and, on a give-and-take principle, in connection with water-works and reservoirs; but as there are also cases in which they may be made, as they have been made, sources of injury, if not even of danger, to the community, their use should certainly never be permitted, except under official sanction and supervision.

CHAPTER VI.

BUILDING SITES.

The bearing of geology upon the important problems of the causes by which health and disease are influenced is briefly referred to in page 52; and the discovery of those causes, with the means of their extension or amelioration, forms the basis of many engineering designs and operations. Sanitary engineering is a modern profession, evolved from the study of such causes, almost during the present generation, and many of the larger works of the present day are designed with reference, not merely to the necessities, but, also, to the health of the community. Such are schemes for the effectual drainage of large towns, the disposal of sewage, and the supply of pure water in quantity sufficient for the requirements of the population. These must all be determined more or less by the physical and geological features of the districts where the works are required. The methods employed in the surveying and proper interpretation of the phenomena presented in the geology of a country have been described in the preceding pages; a few remarks and suggestions follow upon those which are of a purely physical nature, or rather, those which, varying as the geology, may nevertheless be described as surface configuration.

A most important matter, and one which merits more attention than it usually receives is the selection of sites for Public Buildings and for private Residences. On small estates there may not be much choice of situation, but it rarely happens that the limits are so narrow, or the spot so arbitrarily marked out, that nothing can be said on the matter. And even if the exact locality be determined by necessity, or by convenience, there still remains for discussion the questions of aspect, shelter, scenery, and so on. Where there is ample room, and no reason why one spot should be preferred to another (except the desire to select the best possible site) the physical geography is the chief, if not the sole, consideration. For this embraces the distribution of land and water, the geological structure, and the climate of the district; these have a definite relation to each other, and upon them depend all the phenomena by an acquaintance with which a choice should be influenced.

Much of the beauty of a building depends upon the site on which it is placed—that is, upon the fitness and harmony of its immediate surroundings. That which in one spot would strike the beholder as the very type of artistic construction, in another and different locality may seem exactly the reverse. And this is owing, not so much to the kind of material employed, or even to the style of achitecture adopted, as to the planning of the edifice and grounds in regard to their situation.

The term 'site' may have either a restricted or a comprehensive meaning: in the former implying a particular spot in reference to a small area, such as in a plot of ground or on an estate; in the latter indicating some portion of a large district or physical formation. For example, the English National Gallery stands on what has been termed 'the finest site in Europe,' referring to its position in regard to proximate surroundings, on a gentle elevation with commanding approaches from every direction. This is but a small part of tho 'site' of the metropolis itself, which in the midst of a broad flat valley cannot, when viewed on a large scale, be considered 'fine,' notwithstanding its many undoubted advantages.

In the selection of a site in a limited area the position on the map or plan would first be approximately determined with regard to existing roads, drainage, and water-supply, and finally settled with a view to ultimate appearance or elevation. The roads will vary in nearly every instance, and the selection can be subject to no special rules regarding them, but it may be remarked that an important building should never be placed near a road simply for the sake of convenience. The additional privacy to be obtained by placing it at a little distance from the thoroughfare, and the extra facility thus afforded for surrounding it with lawns or shrubberies, are surely worth some little sacrifice on that score, to lay no stress on the consequent improvement in appearance.

The drainage, by which term is meant the natural flow of water, presents a twofold aspect; not only must the getting rid of waste and rain water from a spot be considered, but also the treatment of that which will find its way to it by gravitation. It is not merely by running from a higher to a lower level on the surface of the ground that water abounds in some places more than in others; it is by constant underground percolation that many situations are rendered damp, where dampness would never be suspected from the surface conformation. Questions of site in respect to artificial drainage are at once solved by difference of level—it must simply be somewhat above the point of discharge, whether it be from a main drain, tank, or watercourse; but the natural drainage towards any spot involves a consideration of the geological structure. There may be a water-bearing stratum, a few feet only in thickness, the presence of which cannot be readily detected without special examination. If a house or other building be placed on the outcrop of such a bed the result is perhaps permanent dampness and discomfort therein; if it be but a short distance above or below, the house may be perfectly dry, and the water still made available.

But it is chiefly in regard to water supply that the geological phenomena should be taken into consideration; in this respect also the edifice should be so placed in regard to the water-levels previously described, that pumping may if possible be unnecessary, or where at all events a good portion of the supply may be secured by gravitation.

All the points of plan having been considered, the question of elevation arises, and how this may be effected by the selected position; it may be found that a spot somewhat more to one side or the other will be preferable in this respect, yet possessing the same position relative to certain roads, water supply, and heights for drainage.

Where it can be avoided, houses should never be placed just at the brow of a hill, but so far back from it that the view from the windows does not command all the ground between it and the valley below. Some portion, however slight, should be lost, so that the 'foreground' and 'middle distance' may be definite and distinct, not merged imperceptibly the one into the other; this point is well worthy of remembrance. Any water, and especially ornamental water, ought to be visible from the windows; where nothing of the kind exists, it may frequently, and with little trouble, be obtained. Not by scooping lakes or ponds, but by damming a water-course so that it shall partly fill its valley; this is thought by some not to constitute ornamental water, but the idea seems to be erroneous, for water in a valley looks more natural than on hills and slopes, and there is consequently a fitness about it which adds to, rather than detracts from, real beauty.

For the selection of a site, in its largest sense, that is, a spot in some extended district best suited to the purposes, many points have to be taken into consideration which constitute the details of the district's physical geography. The height above the sea of any spot is, for many reasons, an important element, and one which generally does not receive its due share of attention. The elevation of most places may be readily ascertained with a fair approach to accuracy, for the heights of many points, such as churches, cross-roads, and so on, are figured on the more recent ordnance maps, and the heights of contiguous spots may be estimated by comparison. As a rule, the higher the spot the lower its mean annual temperature, but it does not therefore follow that houses standing on high ground must be cold, or those warm which are built on the lowlands. It is more frequently the reverse, for so many other influences affect the result, although those influences themselves are again partly dependent on the height above the sea. These are, the form of the ground, the direction of the slope (if any), the rainfall, and the dampness of the soil, all partly due to the geological structure. The form of the ground may be either hill, valley, slope, or table-land; if the latter, the spot will probably be warm and dry. A slope should incline towards, or nearly towards, the south; it then receives a full share of sunlight and warmth, and is sheltered by the rising ground in the rear.

A site in a valley may still be at a considerable elevation; such sites are liable perhaps to sudden floods, but dry for a greater part of the year; houses in such situations are well sheltered, unless the direction of the valley meets that of the prevailing wind, and are perhaps as homelike as any that can be found. In the low-lying areas such as broad, flat valleys, a great deal of water is for a time retained, not only the rain which has fallen within them, but that from the high lands in addition, and they are exposed to the full sweep of the wind. For all these reasons an elevated site is much to be preferred to a low one, in all those cases where health, comfort, and appearance can receive their due share of consideration.

The annual rainfall in Great Britain varies from 20 to above 50 inches in depth, its maximum occurring on the west coasts, whence it decreases in an easterly direction, the minimum fall being in Lincolnshire. The rainfall of any intermediate spot is not strictly proportionate to its position between these two points, as local causes produce very great variation. These causes are proximity to, and distance from, the sea, the directions of the principal valleys, and above all others, the height of the hills and mountains. But the rainfall of a district is really no criterion by which we can estimate the climate, unless it be considered in connection with other phenomena. A place may have a very heavy annual rainfall and yet be comparatively dry, or a light one and be nevertheless damp during a great part of the year. The rain may be, as in lower latitudes, very heavy for a short time, a great many inches falling in a few hours; but if the physical configuration be suitable, the water runs quickly away to lower levels and makes but slight local impression. Or it may be that the soils, and the strata beneath them, are highly absorbent, such as chalk and sandy formations, into which the water passes nearly as quickly as it descends. On the other hand, a locality, especially if it be low-lying, may be subject to almost constant drizzling rain, small in amount but persistent, and this produces a damp atmosphere. In respect to the rainfall, therefore, as well as to the form of the ground, those spots are to be preferred as sites which have a considerable elevation; the actual quantity of rain there will probably be greater, but the water will not remain, and the chances of an occasional drought are better than those of perennial saturation.

INDEX.

THE END.

www.ingramcontent.com/pod-product-compliance
Lightning Source LLC
LaVergne TN
LVHW021941220826
846092LV00010B/1199

* 9 7 8 9 3 5 5 2 8 1 9 8 2 *